Leadership and Web 2.0

For Joan,

I'm appreciation of
the generosity and flexibility
& your support for my inquiry

Cathy

Grady McGonagill, Tina Doerffer

Leadership and Web 2.0

The Leadership Implications of the Evolving Web

Leadership Series
BertelsmannStiftung

| Verlag BertelsmannStiftung

Bibliographic information published by the Deutsche Nationalbibliothek

The Deutsche Nationalbibliothek lists this publication in the
Deutsche Nationalbibliografie; detailed bibliographic data
is available on the Internet at http://dnb.d-nb.de.

© 2011 Verlag Bertelsmann Stiftung, Gütersloh
Responsible: Tina Doerffer, Martin Spilker
Copy editor: Tina Doerffer; Josh Ward, Bonn
Production editor: Sabine Reimann
Cover design: Bertelsmann Stiftung
Cover photo: Digital Vision, John Knill
Typesetting and printing: Hans Kock, Buch- und Offsetdruck GmbH, Bielefeld
ISBN 978-3-86793-323-0

www.bertelsmann-stiftung.de/leadership
www.bertelsmann-stiftung.org/publications

Table of Contents

Preface

> *"A door like this has cracked open five or six times*
> *since we got up on our hind legs.*
> *It's the best possible time to be alive,*
> *when almost everything you thought you knew is wrong."*
> Spoken by the character Valentine Coverly in
> Tom Stoppard's play *Arcadia*[1]

Some compare the evolving Web to the revolution of the Gutenberg press. How does the Web shape the role and understanding of leadership? Is a new leadership paradigm emerging? What mindsets, skills and knowledge are necessary for it? What are key challenges, opportunities, pioneering examples and patterns in the business, public and nonprofit sector?

This study aims to enable leaders to anticipate and leverage emerging possibilities in the context of the evolving Web. By providing and analyzing a broad overview of practical examples, it aims to encourage leadership to take new paths as much as to highlight pioneering work. Short examples and longer case studies from Germany, Europe, the United States and other countries are provided that cover all sectors—private, public, nonprofit and a fourth "Commons" sector.

A new leadership paradigm seems to be emerging that is marked by an inexorable shift away from one-way, hierarchical, organization-centric communication toward two-way, network-centric, participatory and collaborative leadership styles. Most of all, in addition to new skills and knowledge, a new mindset seems necessary. All the tools in the world will not change anything if the mindset does not allow and support change. By providing pioneering examples of applications of Web tools, we aim to alter assumptions about the requirements and potential of the Web and to encourage exploration of its possibilities. Such experimentation will ultimately stimulate changes in organizational and societal culture as we adapt to the world of the evolving Web.

Ultimately, organizations and individuals who want to exercise leadership do not have a choice about whether to accept the new world that is emerging. Welcome or not, it is the inevitable future, and is becoming the present in many organizations at a breathtaking pace. The choice is whether and how much to cultivate the culture, mindsets, skills and knowledge that make it possible to leverage the enormous potential of the evolving Web to better realize one's goals. Acknowledging that these changes bring along a number of challenges, it is our belief that, overall, the glass is half-full rather than half-empty.

We would like to thank all who were generous enough to provide us with input and support.[2] And we would like to encourage individuals and organizations to embark on the emerging journey of leadership that is both required and enabled by the evolving Web.

Gütersloh, July 2011

Martin Spilker   *Tina Doerffer*
Director   Project Manager
Bertelsmann Stiftung   Bertelsmann Stiftung

Executive Summary

The Gist of This Study

The culture and tools of the Web are making traditional modes of leadership obsolescent while offering powerful new possibilities for enhancing its impact:

- The Web increases the pressure on leaders in all sectors to become more open and inclusive because it has brought about greater ease of connecting people within and beyond organizations, encourages fuller transparency and has dramatically lowered the costs of collaboration.
- At the same time, the Web offers means of making organizations—and organizing itself—more efficient and effective.

While the Web poses both risks and threats, it also offers enormous opportunities; we see a glass that is more full than empty.

Purpose of the Study

The purpose of this study is to help those who aspire to exercise leadership within—or beyond—organizations of all kinds to understand the implications that the new technologies emerging from the World Wide Web have on their leadership. It does this by:

- Describing in broad terms the impact the Web is having on society and organizations;
- Articulating a new paradigm for leadership that seems most appropriate in light of these patterns of impact;
- Detailing the culture and leadership mindsets, skills and knowledge required by this new paradigm;
- Illustrating via numerous examples the impact the Web is having in each of the three traditional sectors (business, government, society) as well as in a newly emerg-

ing "Commons" sector, which includes social entrepreneurship, multi-stakeholder initiatives and communities, and free-agent leadership;

- Describing how individuals can encourage their organizations to be proactive in exploring whether and how to adapt to the constraints imposed and opportunities posed by the Web;
- Pointing out the known and potential limitations and threats accompanying the Web;
- Describing evolving Web technologies and certain patterns in their evolution.

Overview of Study Findings

Need for a new leadership paradigm: In the past two decades of growing Web impact, the need for a new paradigm for leadership has become more and more apparent. The following seven changes signal that this shift is needed:
- Leadership is now viewed as an activity rather than a role.
- Leadership is now considered a collective phenomenon.
- Individual leaders now need higher levels of personal development.
- There has been a movement away from organization-centric and toward network-centric leadership.
- There has been a shift towards viewing organizations as "organisms" rather than "machines".
- "Learning and adapting" has been replacing "planning and controlling".
- There has been a transition from Generation X to Generation Y.

The paradigm that was dominant until at least the early 1990s assumed that leadership highlighted the dynamic between designated "leaders" and "followers" pursuing shared goals. At its best, this paradigm allowed for participatory and shared leadership, but it inevitably singled out the lone leader as a key player, tacitly reinforcing deeply rooted myths about the importance of "heroic" individual leaders and the effectiveness of "command and control" styles of leading. While situations will continue to exist that are well-suited to this approach, it has become obvious that, in the emerging world, the leadership resulting from this paradigm is increasingly limited in terms of effectiveness.

The need for a new paradigm is rarely disputed, but there is no consensus about what the new paradigm should be like. Indeed, it may very well be that the era of single-paradigm leadership is now behind us. However, what is clear is that the most effective approaches to leadership going forward will be informed by thinking that meets the criteria below (if not others as well). It must be:
- Adaptive (i.e., capable of learning and responding to ongoing change);
- Supportive of emergence (i.e., able to appreciate the fact that systems can spontaneously self-organize and create novel solutions);

- Cognizant of complexity (i.e., aware of the need to bring a degree of input, thought and feeling to challenges commensurate with their complexity);
- Integral (i.e., taking into account a full range of perspectives on people, organizations and society); and
- Outcome-oriented (i.e., more focused on what results from leadership than on the particular ways in which those results are attained).

We describe five illustrative models, each of which meets some or all of these criteria:
- *Action Inquiry* (Argyris 1976; Argyris, Putnam and Smith 1985; Joiner and Josephs 2006; Torbert 1976, 2004), which is a way of simultaneously conducting action and inquiry as a disciplined practice while integrating developmental theory with the skills of individual and organizational learning;
- *Adaptive Leadership* (Heifetz, Linsky and Grashow 2009), which recognizes that leadership is an activity rather than a role, is suited to challenges without known solutions and emphasizes the need for living with disequilibrium;
- *The DAC Model* (Velsor, McCauley and Ruderman 2010; McGuire and Rhodes 2009), which shifts attention away from how designated leaders influence their followers, and instead focuses on the outcomes of leadership (e.g., direction, alignment and commitment, as captured in the acronym) without specifying how those outcomes are created;
- *Integral Leadership,* which is grounded in Ken Wilber's bold aspiration to create a "theory of everything" (Wilber 2001) and aspires to take into account both objective and subjective perspectives on individuals and systems;[3] and
- *Theory U* (Scharmer 2009; Senge et al. 2010), which builds on, deepens and systematizes the best features of organizational learning (Senge 2006) to integrate rigorous data gathering and analysis, deep reflection and practical prototyping of innovations.

Implications for mindsets, skills and knowledge: Assuming that at least some of our readers will be interested in making personal changes and/or helping others make them, we identify the mindsets, skills and knowledge required by new leadership paradigms. The basic shifts in mindset include learning to "let go" as well as shifting the focus from control to influence, from critic to coach, and from heroic leader to facilitator of emergence. The skills that support these mindsets include some traditional ones (e.g., small-group facilitation and individual coaching) supplemented by some newer ones (e.g., self-leadership, developing and leading networks, and convening large groups) as well as some that are tried and true (e.g., interpersonal skills and team facilitation). The knowledge that will support these mindsets and skills includes Web literacy and the appreciation and management of cultural differences, both within organizations and across national cultures.

It will be difficult to nurture qualities of this kind without a supportive culture. We summarize the key features of a culture that is consistent with accommodating the expectations of so-called Millennials and coping with rapidly evolving complexity in the organizational environment, drawing upon the substantial literature pointing to the need for cultures that are more open, transparent and collaborative than is currently the usual case.

Constraints—Radical shifts in the organizational and societal context for leadership: The social media tools of Web 2.0 are shaping the expectations of a new generation of organization members and fostering kinds of interaction and participation that are not only transforming organizations, but also giving them the option to organize themselves. The new media call into question the fundamental value added by organizations while making it easier for individuals to initiate action and form networks on their own. This shift, which sees power moving away from organizations toward networks and individuals, will gain momentum as the Web continues to evolve. However, the Web is already impacting social structures in ways that invite comparison to revolutions as dramatic as the one resulting from the invention of the Gutenberg press. As Clay Shirky (2008) puts it:

> The result is a number of deep, long-term transformations in the culture, structure, process and economics of work. We are shifting from closed and hierarchic workplaces with rigid employment relationships to increasingly self-organized, distributed and collaborative human capital networks that draw knowledge and resources from inside and outside the firm.

Although Shirky assumes a welcoming stance toward these changes, not all observers are as optimistic. They point to threats to privacy, loss of personal space and the dangers of fragmented attention. Regardless of the Web's long-term impact, organizations in all sectors are experiencing effects ranging in magnitude from ripples to tsunami waves as members of a generation that grew up with the Web enter the workforce. The result is pressure to transform existing cultures and traditional hierarchical relationships, challenging both organizations and their managers to become more inclusive in their decision-making and more open and transparent in their operations, both internally and externally. The boundaries around formerly closed organizations are dissolving, as organizations create virtual platforms on which to openly exchange information and perceptions with key players in their evolving environment. On these Web-enabled platforms, organizations that had previously viewed one another as competitors are forming ecosystems of collaboration and mutual support, joined by customers, vendors and other stakeholders.

Opportunities—New tools and modes of exercising leadership: Regarding the second rationale for new leadership paradigms, Web tools offer radical new possibilities for innovation and impact. Web-enabled networks allow organizations to have access to ideas beyond their boundaries while challenging the networks' guardians to define how much of the information previously considered private and proprietary they now wish to share. These tools also offer revolutionary potential for enhanced learning at both the individual and organizational levels. By making a systems perspective on organizations and a global perspective on society more accessible, the Web promises to enable new levels of connection. Indeed, it surmounts the inherent limitations of humans' ability to extend compassion to worlds beyond their immediate experience and to deal with threats that are disconnected from day-to-day realities.

Impact by sector: While this profound cultural shift poses challenges that are common to organizations in all sectors, we see distinctive patterns within each sector. We illustrate these patterns with noteworthy themes, numerous examples and selected case profiles. Below, we provide a brief, sector-specific overview:

- In the *business sector* in particular, the boundaries around enterprises are eroding, enabling deeper and more two-way communication and interaction with and among customers, competitors, suppliers and other stakeholders. Such new constellations constitute "ecosystems" of mutual benefit that are better able to help companies sense and respond to rapidly changing realities. New relationships of this kind, arising from the technologies of "Enterprise 2.0," are better able to meet customer needs while simultaneously drawing customers into the very design of products and services as so-called prosumers, who produce as well as consume. Companies are more able to look for ideas coming from the outside, are becoming more transparent about their aspirations and are drawing upon the best brainpower around the globe. Established businesses also face stiff competition from lean "new industrial era" global players that use the Web to create virtual companies at radically reduced cost and with minimal infrastructure. The new, agile competitors are also able to avail themselves more easily of the economies of "the cloud"—that is, the Internet equivalent of a common, shared resource comparable to an electrical utility—without having to manage legacy IT systems.

- In the *social sector*, individual organizations are increasingly "networked" by using the Web to enhance their effectiveness in attracting support, collaborating with organizations with similar missions and soliciting stakeholder feedback to assess impact. Social media enable self-organizing mobilization to emerge in response to crises and opportunities, requiring established organizations to collaborate more and more with individual "free agents." At the same time, such free agents—acting alone or in networks—are increasingly able to act on behalf of the public good without having organizations as intermediaries. While beneficial for the health of the sector, this trend threatens existing social-sector institutions with obsolescence unless they can demonstrate distinctive value. Nonprofit organizations are also collaborating more with one another in response to greater pressure from funders to produce results as well as to the greater ease of collaboration made possible by the Web.

- In the *government sector*, the Web has breathed new life into "open government" movements in a number of countries across the globe. At all levels of government, agencies in those countries are beginning to make information about their mission and spending more available while seeking information from citizens to better meet public needs. Public bureaucracies are becoming more transparent about

15

their operations and decisions not only to the public, but also to their employees and other agencies. Indeed, government is acting more like business, treating the public as customers to be served and holding itself more accountable for meeting the needs those customers are now better able to articulate. To this end, government institutions are increasingly forming "policy webs," in which a wide range of stakeholders participate in the decision-making process. Examples are emerging that support the vision of Web tools as enablers of more effective decision-making within bureaucracies and as a means of linking formerly unconnected or marginalized citizens in ways that facilitate "emergent democracy."

- Increasingly, individuals and organizations are called upon to come together across sectoral boundaries and to find common cause in the effort to address "wicked" problems defying solutions from within any single sector. This report points to the emergence of what we are inclined to call a *"Global Commons."* This new Commons has a number of discrete ingredients, all of which serve to enhance the well-being of the collective. It is a critically important source of the new leadership's ability to address "stuck" problems at all levels. This fourth "sector" includes: a range of cross-sector, "blended" initiatives that reflect the goals of civil society and/ or government while using the mechanisms of both business and business enterprises that have adopted explicit non-monetary goals to address social values. It includes multi-sector efforts to collaboratively address challenging problems that cannot be solved by any single sector and to form "megacommunities" of ongoing relationships, and it also contains the rapidly emerging phenomenon of leadership by individuals acting as "free agents." We see this sector as continuing to become more and more significant, eventually to a large extent subsuming the discrete sectors as people within, across and/or outside organizations rise to the challenge of collaboratively constructing sustainable lifestyles, cultures and societies in a world of increasing complexity, accelerating change and daunting problems.

The art of letting go: Thus, Web tools and the culture they bring with them pose both challenges and opportunities for those who would lead, whether within or outside of organizational settings. For better and/or worse, these tools tend to make information freer, organizations more transparent and boundaries more fluid. In the world that is emerging, strong pressures and rewards combine to encourage institutions to make accommodations for these new realities. Managers in all sectors and at all levels must learn "the art of letting go"—that is, relinquishing the control they feel they need, but may in fact have lost long ago. Similarly, there are strong pressures on managers to exercise leadership in ways that are more open, inclusive and participative.

Options for organizational adaptation: In response to these trends, organizations are being challenged to figure out how to position themselves. They are being forced to ask themselves questions, such as: How open do we wish to be in terms of sharing information and expanding participation in decision-making? How can we foster the needed changes in our organizational culture? This study suggests that the question of which Web technologies one should embrace is secondary to the question of how to clarify the organization's desired stance toward the new culture more generally, and particularly in light of the organization's objectives. At the most basic level, this is a question about culture, not technology. Only when organizations choose their preferred stance and embrace the Web's implications for cultural change does it make sense to decide whether to use particular social-media tools. Indeed, rather than being a question of which tools to adopt, it is primarily one of appreciating—and, for most organizations and most managers, to some degree adopting—the shift in mindsets that accompany the cultural transformation catalyzed by the Web. Individuals aspiring to leadership outside an organizational base can ask themselves a parallel set of questions about how they wish to communicate with and mobilize others in response to their individual voices when prompted by inspiration or chance opportunities.

On the assumption that many organizations will elect to take at least tentative steps toward the Web and its culture, this study goes on to offer a few tips on how to do so. It is safe to predict that—for most, if not all organizations—desired change will bump up against both organizational and individual "immunities to change." Organizational culture and the individual habits of thought and action that support it are very difficult to alter. To do so, you must take these two important steps: First, you must recognize the naturalness and legitimacy of what may appear in the organization, in others or in oneself as "resistance." And, second, you must set clear priorities regarding what changes—in culture or in one's own mindset—make sense and offer the most leverage given your particular organization and personal approach.

This study suggests seven steps that individual managers (or others) can take in organizations to encourage a strategic approach to adapting to a new culture of transparency, openness, interaction and collaboration. Specifically, we recommend that managers:

1. Gain personal Web literacy and encourage members of their team to do so as well;
2. Encourage a strategic planning process that addresses Web strategies;
3. Encourage development of organizational policies regarding the use of social media;
4. Encourage someone in the C-suite of their organization to start a blog;
5. Encourage your human resources, marketing and communications departments to experiment with social media;
6. Help the organization anticipate common barriers and pitfalls of adopting Web tools; and
7. Ensure development of Web strategies from multiple perspectives.

Liabilities of the Web: Of course, the enormous cultural shifts in society and organizations do not come without costs and risks. Although they vary somewhat by sector, a common set of risks and threats encountered by organizations—and, thus, their leadership—includes at least the following:

- Concerns about the validity and security of information and the danger of being overwhelmed by a swelling volume of data;

- A fear that the multitasking made easy by the Internet will lead to a decline in functional intelligence, the quality of consciousness and business productivity, as well as decrease personal space and time;
- An unrealistic faith in the superior "wisdom" of computers combined with the failure to recognize the limits of collective intelligence;
- Threats to authority and genuine expertise in an era in which everyone has a voice;
- A "digital divide" that results from uneven access to technology and could exacerbate the gap between the haves and have-nots.

In an appendix, we review trends in the evolution of the Web. We do so by incorporating a popular scheme for distinguishing among three phases in the evolution of Web technologies while at the same time pointing to the limits of such a neat categorization:

- *Web 1.0* (1991–2000), in which tools for faster, cheaper and more convenient forms of communication (e.g., e-mail) became widely available and used.
- *Web 2.0* (2001–2010), in which use of another set of new tools for communication (e.g., wikis and blogs) began enabling interaction and communication in transformative ways.
- *Web 3.0* (2011–present), in which powerful new computing platforms (e.g., "the cloud"), a second generation of search tools and meta-level methods for managing knowledge (e.g., tags and "folksonomies") are beginning to realize the Web's potential to generate more immediately and personally useful knowledge from archived information.

Conclusion: In sum, we argue that organizations and those who intend to exercise leadership have no choice about whether to accept a new world that is fundamentally different from the old one. Welcome or not, it is the inevitable future, and is becoming the present in many organizations at a breathtaking pace. At the same time, there is a choice about whether to deny or react against these cultural and economic shifts, or instead to acknowledge and embrace them. There is a choice as well—for both organizations and individuals—about whether and how much to cultivate the culture, mindsets, skills and knowledge that make it possible to leverage the enormous potential of the tools of the evolving Web to better achieve their purposes.

It was true before the Web, and is even more true as a result of it, that most of us are usually "in over our heads" in relation to the challenges we face. We are also facing an increasing number and variety of threats to our security, privacy and peace of mind. However, thanks to the Web, we also have an opportunity to learn how to hone and extend our individual intelligence, deepen our collective intelligence and use the Web's tools to address the threats to our well-being and survival that have resulted from the accumulated, unintended systemic consequences of our behavior. Thus, the ultimate implication of the Web for leadership is that it provides hope for a sustainable future—and the tools to help create it.

1 The Nature of the Study

Purpose and Rationale

This study aims to make explicit the implications for leadership of the new technologies made available through the Web. It aspires to do so by considering simultaneously co-evolving trends in society and in our understanding of the world in which leadership is exercised. Looking at these combined trends will enable people in leadership positions within organizations across all sectors to better anticipate and prepare for emerging opportunities and threats.

Two factors motivate the Bertelsmann Stiftung to sponsor this study:

- Social and technological trends are having a revolutionary impact on organizations and organizing across the world and have powerful implications for the understanding and practice of leadership.
- Understanding and anticipating these implications provides a high-leverage opportunity for leaders in all sectors to enhance their effectiveness.

Methodological Assumptions and Limits

To accomplish its purpose, this study will:

- Map and analyze patterns and trends in the evolving Web;
- Summarize key trends in our understanding that affect organizations and leadership;
- Identify specific implications for leadership and its development in all sectors.

The study has been undertaken under significant constraints on resources and time. Our principal research methods have been:

- Identifying relevant literature and Web-based information;
- Inviting input from and selectively interviewing nearly 40 colleagues and experts in the field; and
- Interviewing more than a dozen practitioners associated with programs that were brought to our attention by the sources mentioned above.

2 The Implications of the Web for Leadership

The Implications of the Web for the Context of Leadership

A revolution with few precedents

In order to understand the implications of Web technologies for leadership, it is important to first understand the enormous changes that are evolving from these tools within organizations and in the nature of organizing itself. Social media are shaping the expectations of a new generation of organization members and are fostering patterns of interaction and participation that are transforming internal organizational structures, processes and relationships, as well as external relationships with customers, competitors, suppliers and other stakeholders. The new media call into question the fundamental value added by organizations as they make it easier for individuals to initiate action and form networks on their own.

Although this shift will gain momentum as the Web continues to evolve, the Web is already impacting social structures in ways that invite comparison to inventions with impacts as dramatic as that of the Gutenberg press. According to Clay Shirky (2008):

> The result is a number of deep, long-term transformations in the culture, structure, process and economics of work. We are shifting from closed and hierarchic workplaces with rigid employment relationships to increasingly self-organized, distributed and collaborative human capital networks that draw knowledge and resources from inside and outside the firm.

Is the glass half full or half empty?

Shirky assumes a welcoming stance toward these changes, as do other prominent voices. Former *Whole Earth Catalog* editor Kevin Kelly welcomes the emergence of a "global mind" (Kelly 1994: 202), as do *MacroWikinomics* coauthor Don Tapscott (Michalski 2008) and German network guru Peter Kruse (Kruse n.d., 2010). Some enthusiasts go so far as to assert that "the network, patterning structure of what a mind can know is mirrored in the network, patterned structure of the Open Internet," which leads to the conclusion that "what is known by humankind has spontaneously nestled into the Internet and begun interconnecting itself there, as an embedded cognitive network" (Breck 2005: 1).

Still, not all observers are as optimistic. Instead of using the analogy of the Gutenberg press, some liken the Web to the Bolshevik Revolution, which promised a utopia but delivered a nightmare.[4] These skeptics and others point to threats to privacy and national security, the concern that having our attention fragmented is causing us to become stupider rather than smarter (Carr 2010), and the fear that, while giving us access to collective wisdom, the Web is also making us more vulnerable to "hive mind"—that is, collective folly (Lanier 2006, 2010).[5] Whether one sympathizes with the optimists or the pessimists, it is undeniable that radical change is coming.

The "tectonic shift" catalyzed by Web-based technology is not limited to any one sector, but will impact them all. To be sure, organizations will retain some of their traditional forms; corporations, governments and foundations will not go away. But the relative advantage of such forms of organization has disappeared. As Shirky puts it (Shirky 2008: 24): "The new possibilities for self-organizing, group communicating, sharing and action will transform the world everywhere (that) groups of people come together to accomplish something, which is to say everywhere."

The future is already here

Estimates of the level of usage of Web tools evolve rapidly, shifting even between drafts of this study. Nevertheless, the overall pattern is clear: Whether we like it or not, the Web is coming. Just how visible it is depends on where you are located—which country, which part of that country—and which kind of organizations and networks are part of your life. This brings to mind novelist William Gibson's famous observation: "The future is already here—it's just not evenly distributed."[6]

There are a number of common patterns of Web impact across sectors. These include greater openness, transparency, participation and collaboration (resulting in part from lowered costs). In Chapter 3, we review the impact of the Web by sector using many specific examples. But, for now, we wish to give an overview of the leadership implications of this impact.

The Need for a New Leadership Paradigm

The impact of the Web on leadership is evident in two ways: It requires people in positions of formal authority to think and act differently because of the way it is changing their external and internal organizational environment, and it simultaneously provides them with new opportunities for leading and learning. Trends in society that have overlapped with the evolution of the Web have also fostered fundamental shifts in ways of thinking about leadership. As a consequence of these simultaneous trends, we see a steady movement away from an old paradigm, which features leaders and followers in relation to goals, toward a new way of thinking, which is more focused on the desired outcomes of leadership than on how it is achieved.[7]

Seven Indicators of the Need for a New Paradigm

In surveying the vast literature on leadership in recent decades, we see the following seven trends, which—taken together—suggest we need new mental models for leadership.

Leadership as an activity rather than a role

It is more useful to view leadership as an activity than as a position. Ronald Heifetz has long encouraged distinguishing between leadership as a role and as a behavior (1998). From this perspective, people are leaders not by virtue of their role but because of what they do in any role. They are persons at any level within or outside organizations who are "actively involved in the process of producing direction, alignment and commitment" (McCauley and Van Velsor 2004: 2). Although there remains an important leadership role for those in positions of formal authority, it can no longer be assumed to be the role of the "heroic" leader who shoulders both responsibility and control. In most positions, and in most organizations, this role must shift in the direction of "post-heroic leadership," which accesses the wisdom and releases the potential of others (Bradford and Cohen 1998). As Haeckel puts it (Haeckel 1999: 93): "Context and coordination replace command and control."

The Web and other trends have accelerated the obsolescence of "command and control" as the default leadership style, and have increased its risks. To avoid these risks and begin the shift toward a new paradigm, Web 2.0 requires leaders to cultivate the "art of letting go" (Buhse and Stamer 2008). The new challenge of leadership is to foster the best efforts of individual contributors and to nurture the emergence of the broadest possibilities from the collective.[8]

Leadership as a collective process

James MacGregor Burns, often considered the father of the leadership development field, and author of the seminal book *Leadership* (1978), was asked in an interview about the next frontier for the field of leadership. Without hesitation he answered: "We need to better understand leadership as a collective process."[9] Thinking of leadership as an activity rather than as a role exercised by designated leaders makes it easier to consider the possibility of collective leadership. Shifting emphasis from individual leaders to the interaction among leaders and followers leads to an appreciation of the possibility that leadership can spontaneously emerge from the collective (Hubbard 2005). This observation echoes a trend in practice away from a focus on individual formal leaders to the interaction among leaders and followers and the spontaneous emergence of leadership from the collective (ibid.). As Drath and Palus (1994) put it: "Leadership is … about creating a 'system' or 'culture' in which members instinctively do the 'right thing' even when the official leaders are absent." Or, as Haeckel puts it (Haeckel 1999: 93): "In an environment of discontinuous change … leaders can no longer know as well as followers how to get things done."

Collective leadership recognizes that wisdom can reside within a group. Under certain conditions, groups have been shown to generate more accurate information than experts and to make better decisions than individuals (Surowiecki 2004). Likewise, there is mounting support for the idea that an intelligence can sometimes emerge in groups that transcends the intelligence of its individual members.[10] Although Burns' comment is relevant to all sectors, the social sector has been a pioneer in this area over the last decade and has provided many robust examples.[11]

It has often been possible for the "great man" (and, more recently, "great woman") to emerge as a leader because of some combination of motivation and charisma. However, it is now becoming possible for ordinary people to exercise leadership—or at least contribute to it—on a much more spontaneous and temporary basis. For example, individual bloggers in Thailand who posted photographs they took of tanks in front of the parliament building exercised a form of leadership in documenting the 2006 military coup. More recently, anonymous protesters on the streets of Iran were able to catalyze sympathy and support across the world by spontaneously taking a photo with a cell phone and uploading it to YouTube.[12] Some of these "leaders" seem highly improbable candidates to play such a role—e.g., "a fashion-obsessed college student" (Shirky 2008: 37).

Need for individual leaders at higher levels of development

Over the past several decades, a particular set of theories on human development has had an increasing influence on leadership theory. Constructive-developmental psychology posits that human development does not stop when we reach adulthood, but rather continues (or has the potential to continue) throughout one's life (Kegan 1982; Torbert et al. 2004). According to this view, people evolve through stages, and leaders who have attained higher stages of development are able to draw on greater complexity in the way they see the world and the options they can imagine and implement (Joiner and Josephs 2006). The title of one of the seminal books in this field captures the importance of leaders who have such qualities: As Robert Kegan (1994) puts it, we are all "in over our heads." In the face of the increasingly daunting challenges we face, we need the help of people who have the capacity to make sense of the systemic complexity of problems. The good news is that such people exist; the bad news is that they are rare. Research suggests that fewer than 10 percent of managers have evolved to the optimal range of capacities (Torbert et al. 2004). However, we are learning about approaches to leadership development that can foster such development.[13] Learning how to import those methods into mainstream organizations has an increasingly high priority.

From organization-centric to network-centric leadership

Networks are increasingly being recognized as the way things actually get done within organizations (Cross and Thomas 2008). The personal networks of leaders and networks among team members are recognized as critical to implementing leadership initiatives. For example, the U.S. military has discovered that networks are the best way to respond nimbly to a rapidly changing environment, because they allow information, technology and combat assets to be used as efficiently as possible (U.S. Department of Defense 2001). Understanding how networks function can also increase leadership effectiveness in getting things done (Anklam 2007: 226). In addition, the emerging need to address complex problems that defy the abilities of any single organization or sector has led people to recognize the importance of building networks to forge relationships. Such relationships serve as a source of leadership across organizational and sectoral boundaries, as well (Fine 2006: 50–51).

Web 2.0 is making networks more prevalent and more powerful. Indeed, understanding networking is increasingly inseparable from understanding Web technologies. The power of the Web to support networks extended in space and time will make skills in this area even more critical. In an influential article in *Foreign Affairs*, Anne-Marie Slaughter argues that as its financial dominance wanes, "America's edge" in the emerging world will derive from how it exercises influence through global networks (Slaughter 2009).

25

From organizations as "machines" to organizations as "organisms"

Although networks have been around for millennia, network theory arose only a little more than a decade ago from within the field of complexity—one of the "new sciences" relevant to leadership heralded by Margaret Wheatley (1992). She suggested that it no longer made sense to try to understand organizations as if they were "machines," as suggested by the Newtonian paradigm.[14] To her, it was becoming clear that it makes more sense to view organizations as "organisms." In the two decades since, the power of this new metaphor has become increasingly evident. Similarly, the power of two other concepts related to the new science of complexity—"emergence" and "self-organizing"—are also now obvious (Holland 1996). Many examples of emergent self-leadership have become visible. Companies as diverse as IBM, Eli Lilly, Harley-Davidson, Procter & Gamble and dozens more "are discovering how to make self-organization a key component of the modern-day strategic arsenal" (Ticoll and Hood 2005).

Emergence and self-organization are key features of Internet culture, which has been profoundly shaped by the open source software movement. The "geek culture" brought with it new images of what it means to develop a product and bring it to market. Eric Raymond, a member of that culture, reported the shift that he saw and personally experienced. In his view, the dominant approach had been governed by the metaphor of the "cathedral," in which something is carefully crafted using a centralized, coordinated approach until it is as perfect as possible. By contrast, development of Linux proved the power of a different metaphor, "a great, babbling bazaar of different agendas and approaches … out of which a coherent and stable system could seemingly emerge … by a succession of miracles" (Raymond 1999: 21–22). According to this new approach, "wizards" working alone or in small bands are replaced by lots of users treated as if they were experts. As Raymond describes it, "many eyeballs tame complexity" and "given enough eyeballs, all bugs are shallow." This movement has contributed to a fundamental shift in mindsets and metaphors away from static, rigid structures (e.g., machines, cathedrals) toward more fluid and dynamic ones (e.g., organisms, bazaars).

From planning and controlling to learning and adapting

Technological change is, of course, by no means new. But there is something unique about the speed with which the emergence of the Web is taking place. In the United States, we are seeing "an adoption rate for the digital infrastructure that is two to five times faster than adoption rates were for previous infrastructures, such as electricity and telephone networks" (Hagel, Brown and Davison 2010: 48). The implications affect us all. "Until recently," they write, "one could notice something emerging on the edge and—because it would take so long for its effect to be felt in the core—safely ignore it. … We are now in a different era, one where edges emerge and rise up with astonishing speed to catalyze changes on a global basis in less time than ever before" (ibid.: 57). The accelerating rate of change means that traditional tools for making sense of the world and planning action are breaking down. A manager in a major engineering firm said: "We don't plan anymore. ... Given the pace of change, why bother planning? You see what's coming down the pike, and you go with it" (Gerencser et al. 2009: 205). Indeed, this has become the norm. Stephen Haeckel points to the "disbandment of large central planning departments" (Haeckel 1999: 11). Thus, he writes, "the only kind of strategy that makes sense in the face of unpredictable change is a strategy to become adaptive" (ibid.: xvii). The implication of this and many other indicators is that leadership itself must become adaptive.

Adapting to these new realities will certainly require learning. Happily, our ability to learn is enhanced by the Web. In fact, one of its more exciting and potentially profound potentials is its ability to enable individual and organizational learning. At the individual level, this has accelerated the trend toward "blended learning," in which classroom instruction becomes just one of many modes of learning, providing support over time and when it is most needed, rather than concentrating it in large doses that are of questionable long-term impact (Schooley 2009). At the organizational level, this creates new mechanisms for peer-to-peer learning through the sharing of best practices, for getting feedback from internal and external stakeholders of all kinds, and for more self-initiated learning (Haeckel 1999: 82). At the same time, discontinuous change requires learning that is faster and deeper than ever before. Not only does it require adaption to a given context, "it requires adaptation of the context itself." When operating "in environments of discontinuous change, thinking outside the box is not sufficient: It is also necessary to think about changing the box" (ibid.: 82).

From Generation X to Generation Y

Although technology is the enabler, the real driver of change is cultural, and one of the strongest forces driving this change is the arrival of a generation in the workplace that has "grown up digital" (Tapscott 2008c). These former "screenagers" are referred to variously as the "Net Generation" (ibid.), "Generation Y" (Erickson 2009), "Generation F" (for Facebook) (Hamel 2009) and "Millennials," that is, those born roughly in the last two decades of the 20th century (Lancaster and Stillman 2002; Meister and Willyerd 2010).[15] Whatever the name, as of 2010, this generation constitutes a portion of the workforce equal to that of baby boomers in developed nations, and in fact outnumbers baby boomers globally (2.3 billion versus 1.4 billion).[16] This generation brings with it expectations from the culture of the Internet.

The culture that has shaped the mindsets and habits of Millennials has been strongly influenced by the "geek" culture of free and open software, which places high value on freedom and the open sharing of information. Some commentators see a "new socialism" evolving from this culture, which will clash with norms that evolved from capitalism (Kelly 2010). But the trend will surely aid the ability of organizations to survive in turbulent times. As Bernholz (2010) puts it: "High levels of adaptive capacity are typically achieved through … the free flow of communication and ideas, especially between and across different levels, e.g., bottom-up and top-down."

The Web-based tools to which this generation is accustomed enable easy sharing of information across traditional boundaries, which will inherently encourage more feedback from stakeholders of all kinds: employees, customers and even the public at large. David Eaves (2010) suggests that this gives rise to an opportunity to create "patch cultures." He borrows the metaphor from the open source software community, in which members hold the shared assumption of an ongoing need to correct inevitable errors—in other words, to "patch" them. In such a culture, error is not viewed as a sign of incompetence, but rather as the necessary consequence of improvisation, experimentation and shared responsibility. The expectation and acceptance of error is simply part of the new landscape. In a world of constant flux, survival means adaptation, and no person or organization can adapt without making mistakes. The value of such a culture transcends national boundaries. An interview with a senior project manager in this study's sponsoring organization, the Bertelsmann Stiftung, reminded us of the importance of a "Fehlerkultur" (literally, a "mistake culture") if decision-making is to be more widely distributed. "If you delegate to the bottom," the project manager said, "you need to accept some errors."

The arrival of this generation will create discomfort for people in leadership roles who are used to maintaining control over information and limiting the feedback they receive (Weinberger 2008). And for organizations that try to keep pace with the new culture, it will test whether they are able to avoid the gap that often emerges between an organization's "espoused theory" and its "theory-in-use" (Argyris, Putnam and Smith 1985).

Criteria for a New Paradigm

We believe that, when taken together, these signs constitute a compelling case for a new leadership paradigm (or perhaps even more than one). Indeed, it may be that the era of single-paradigm leadership is now behind us. Attractive as it is to identify the next new paradigm, we think it is more realistic to view the current situation as one of intense fermentation. We seem to be living in a period of continuous disequilibrium, at the boundary between order and chaos, which complexity theory teaches us is the most fertile ground for creativity.

What is clear is that the most effective approaches to leadership going forward will meet criteria, such as being:
- Adaptive (i.e., capable of learning and responding to ongoing change);
- Supportive of emergence (e.g., able to appreciate the fact that systems can spontaneously self-organize and create novel solutions);

- Cognizant of complexity (i.e., aware of the need to bring a degree of input, thought and feeling to challenges commensurate with their complexity);
- Integral (i.e., taking into account a full range of perspectives on people, organizations and society); and
- Outcome-oriented (i.e., more focused on what results from leadership than on the particular ways in which those results are attained).

A Sampling of Alternative Paradigms

Below, we describe five illustrative models that we find attractive. Each of them meets some or all of the above-mentioned criteria:

Action Inquiry (Torbert et al. 2004; Joiner and Josephs 2006): Action Inquiry is a way of simultaneously conducting action and inquiry as a disciplined practice. It is a "meta-theory" in that it integrates multiple paradigms. One of its two principal ingredients is an elaboration of the "action science" paradigm first named by Torbert (1976) and further developed by Argyris, Schön and others (Argyris, Putnam and Smith 1985). It is based on evidence that people in all cultures unknowingly tend to act according to a set of tacit governing values that are unilateral and self-protective while at the same time espousing collaborative and open bilateral values. This leads them to be unaware of their unwitting contribution to patterns of dysfunction of which they perceive themselves to be the victim. Furthermore, they tend to be blind to ways in which they do not act consistently with what they preach, and consequently lose credibility. Andrew McAfee, who is credited with coining the term "Enterprise 2.0" (i.e., applications of Web 2.0 to business), acknowledged the relevance of this model in his recent book on that topic (McAfee 2009a).

The other principal ingredient is developmental theory. Torbert has developed his own elegant stage theory of human development by building on the work of others (Erikson 1994; Loevinger 1976; Kegan 1982, 1994, 1998). Torbert's colleague and student Bill Joiner has refined the developmental theory and applied it explicitly to leadership in *Leadership Agility* (2006), a book that is both conceptually powerful and eminently practical. One chapter summarizes how a hypothetical manager, Ed, would cope with the same situations from the vantage point of five different stages of development. As Ed moves into the fourth and fifth stages, he begins to frame problems more broadly and conceive of solutions that are increasingly complex and creative.

Adaptive Leadership (Heifetz 1998; Heifetz, Linsky and Grashow 2009): Heifetz (1998) defines leadership as "mobilizing people to tackle tough problems." This means addressing "adaptive" problems as opposed to "technical" ones. This approach contrasts with the notion that leaders should have a vision and align people with it. Here, solutions lie in collective wisdom rather than in leaders' minds. As already mentioned, Heifetz frames leadership as an activity, not a position. We are leaders only to the extent that we act, and this can be done from any organizational (or societal) role. He describes the widespread and "maladaptive" tendency to seek solutions among people in authority. Heifetz draws a wide range of examples from the worlds of politics, business and medicine. In a related article, Heifetz and Donald Laurie (2001) offer six principles: "get on the balcony" (i.e., step back from the field of action to see the context); "identify the adaptive challenges" (i.e., pinpoint how an organization's value systems or methods of collaboration need to change); "regulate the inevitable distress" (i.e., contain anxiety); "maintain disciplined attention" (i.e., address differences in employee habits and beliefs); "give the work back to people" (i.e., let employees take initiative); and "protect the voices of leadership" (i.e., encourage voices from below). His most recent book (2009), written with two colleagues, applies these principles in practical ways to leadership challenges of the complexity typical of today's turbulent environment.

The DAC Model (Velsor, McCauley and Ruderman 2010; McGuire and Rhodes 2009): For well over a decade, scholars and practitioners associated with the renowned Center for Creative Leadership (CCL) have been pioneering a systematic effort to articulate a new paradigm (McCauley and Brutus 1998; McCauley and Van Velsor 2004; Drath et al. 2008; McGuire and Rhodes 2009). They point out the limits of the existing paradigm tacitly underlying most of the previously dominant definitions, which focuses on the relationship between leaders and followers in pursuit of shared goals. They argue that this paradigm is a special case of a more robust, outcome-oriented paradigm, which leaves open how those outcomes are to be attained. In this view, the purpose of leadership is to ensure three outcomes: direction, alignment and commitment.

This shift in emphasis—from the process of leadership to its results—throws into sharp relief the weakness of the traditional emphasis on the contribution of individual leaders. It underscores the need to also pay attention to leadership capacity, defined as "the organization's capacity to enact the basic leadership tasks needed for collective work: setting direction, creating alignment, maintaining commitment" (McCauley and Van Velsor 2004). Granted, these outcomes can certainly result from the actions of individual leaders in positions of authority who interact with followers in pursuit of mutually agreed-upon goals. But they can also result from a variety of other interactions that result less directly from the actions of such leaders, including spontaneous initiatives from people who are not in positions of formal leadership but who nevertheless mobilize others.

This paradigm recognizes that different parts of an organization can play different but complementary leadership roles to help the complex adaptive system of the organization position itself for survival. Accordingly, it would allow both hierarchy (for executing clearly defined objectives) and networks (for exploration, collaboration and innovation) to coexist.[17]

Integral Leadership (Volckmann 2010; Wilber 2000): Ken Wilber has an enormous appetite for reading and an unusual talent for synthesizing what he has read. Over several decades, he has synthesized readings that encompass Eastern wisdom as thoroughly as Western science and philosophy. The result is an "integral" way of thinking

30

about change in individuals and systems at all levels. Wilber makes the generous assumption that all schools of thought—in all domains—have some merit. In his view, all approaches offer some truth, though it is always only a partial truth. The challenge is simply to find out their limits—that is, the conditions under which their "truth" obtains. Thus, he aims to construct a meta-theory that identifies the distinctive contribution of as many theories as possible.

The resulting framework brings an "integral" way of thinking to any given topic. Wilber's framework, often referred to by the acronym "AQAL" (all quadrants, all levels) synthesizes wide-ranging sets of theories about levels of development, lines of development, personality types, states of consciousness and "quadrants" (Wilbur uses a useful 2x2 matrix that shows how internal/external and individual/collective axes create four different lenses through which to view any given situation). This theory has attracted many theorists and practitioners, who are applying the framework to a variety of areas, including leadership (McIntosh 2007). The broad foundation of this work makes it a truly "meta" theory, in that it self-consciously strives to integrate all other theories into a "theory of everything" (the title of one of Wilber's books). Although the definitive book applying this theory to leadership has yet to be written,[18] the work of Joiner and Josephs (mentioned above in connection with Action Inquiry) embodies many qualities of the integral model. However, it does not attempt to comprehensively cover all aspects of Wilber's theory.

Theory U (Senge et al. 2005; Scharmer 2009): Otto Scharmer's compelling theory represents a major, path-breaking contribution to leadership theory and practice. It develops in much greater depth the "U Theory" described by Scharmer and co-authors Peter Senge, Joseph Jaworski and Betty Sue Flower in *Presence* (2005). It resulted from interviews Jaworski and Scharmer conducted to understand how to help leaders learn better how to sense what was needed in the world and bring it forth—that is, how to "learn from the future." They interviewed over 150 thought leaders from around the world in the areas of creativity, high performance and leadership, including economists, entrepreneurs, cognitive scientists, educators and Eastern gurus. These interviews persuaded them that leaders will have to address a "blind spot" in our understanding of leadership. What's more, leaders need to develop a new cognitive capacity that involves "primary knowing," that is, knowing in a more holistic and intuitive way. From this perspective, "the most important tool for leading 21st-century change is the leader's self." From this perspective, cultivating and deepening one's personal qualities offers the greatest leverage for increasing one's impact as a leader. "Theory U" is Scharmer's articulation of the process whereby leaders can move from taking in new information to accessing their capacity for making deeper sense of that information and to subsequently envisioning and prototyping new innovations. The theory depicts three spaces that are envisioned in the form of a "U": sensing (i.e., seeing current reality), presencing (i.e., reflecting deeply) and realizing (i.e., acting). A robust, global community of practitioners has been attracted to Scharmer's Presencing Institute.

These choices undoubtedly reflect our own mindsets and biases. We intentionally excluded some excellent models because we judged them as still being grounded in the old paradigm (e.g., Bradford and Cohen 1998). We excluded other models because we were not able to become sufficiently familiar with them to be confident of passing reasonable judgments (e.g., Stacey 2010). Whatever the limitation of any particular choices, we believe that, as a whole, the set we provide ably illustrates the emerging landscape of possibilities.

Implications of a Paradigm Shift

The Web technologies that have coevolved with societal trends will increasingly serve as nails in the coffin of the old paradigm, while accelerating and consolidating the emergence of a new one. Thus, we can expect more open and participative forms of leadership to play an increasingly important role. Web 2.0 expands both the capacity and the disposition of people throughout an organization to communicate with one another and to form links with people outside the organization, whether they are customers, suppliers or peers (Li 2010).

However, any new paradigm would do well to leave room for clearly defined and traditional leader/follower roles, with even the "command and control" variant of the "heroic leader" approach being honored as a special case. New paradigms simply expand the space of possibility and encourage a strategic choice of style. It is nonetheless clear that more traditional styles will become ever riskier in light of the need to understand and adapt to a rapidly evolving environment. As discussed in depth in Chapter 3, businesses must cope with a world that is increasingly interdependent, hypercompetitive and characterized by an accelerating rate of change. Traditions in organizational practice and leadership that limit learning from the environment and responding flexibly to it will not only be unattractive to the Millennials, who constitute the next wave of membership, but also threaten an organization's very survival.

Developing New Leadership Mindsets, Skills and Knowledge

Whatever the particular realities of any given organization, it is safe to assume that most people in roles of formal authority across all sectors will need to develop new mindsets and skills in order to master the kinds of leadership most effective under the evolving conditions. For the few people in these roles that don't have such needs, it is surely the case that they will need to help others develop those new capabilities.

Mindsets

As a new world evolves with and as a result of the Web, the foundation for the leadership required in it is the way leaders think—that is, their mindsets. Likewise, culture is primarily defined by the underlying and often tacit assumptions of an organization's members—again, their mindsets. Thus, it is critical that the underlying assumptions that drive their behavior be consonant with the desired organizational culture. Reporting on interactions with hundreds of leaders discussing the power of social technologies, as well as on research on many organizations that have taken the plunge, Li writes that (Li 2010: 8): "The biggest indicator of success has been an open mindset—the ability of leaders to let go of control." Indeed, the "art of letting go" is often identified as the core mindset shift critical to riding the wave of the new culture (Buhse and Stamer 2008; Weinberger 2008). The success of the Obama campaign illustrated this mindset in a dramatic way (DiJulio and Wood 2009). For example, he "let go" of directly managing the development of his foreign policy platform, which enabled him to draw upon over 300 groups through distributed leadership. By contrast, McCain was able to get advice from only a quarter of this number through his more direct oversight, using a comfortable but woefully ineffective "command and control" style. This is not a revolutionary message. Practitioners on the cutting edge of the old leadership paradigm learned some time ago that it is both more realistic and more effective to focus on influence rather than on control, as well as to frame influence as being mutual rather than unilateral. A related shift of mindset is from ROI to ROR—"return on investment" to "return on relationships." Since things get done through people, it is necessary to build relationships with peers and others over whom one has no authority (Li 2010: 9). These and many other mindset shifts are necessary in order to capitalize fully on the potential of the Web.

Such shifts in mindset are increasingly necessary rather than merely optional. Indeed, by 2020, Meister and Willyerd predict that employees (Meister and Willyerd 2010: 222–223):

> … will communicate, connect and collaborate with one another around the globe using the latest forms of social media. As they work in virtual teams with colleagues and collaborate with their peers to solve problems and propose new ideas for business, they will need to develop a new mindset to thrive. The 2020 mindset will incorporate abilities in:
> * Social participation …
> * Thinking globally …
> * Ubiquitous learning …
> * Thinking big, acting fast and constantly improving …
> * Cross-cultural power.

Of course, we have also learned that mindsets are hard to change (Kegan and Lahey 2009) and that they tend to be embedded in a variety of ways. Despite the implications of "mind" as being cognitive, there are often emotional dimensions reinforced by neural pathways in the brain (Goleman 2006). Nevertheless, the place to begin is at the cognitive level. First and foremost, leaders need to have a deep understanding of how the paradigm is changing. This will mean "unlearning" old mindsets about leadership as well as unlearning the reflexive default behaviors that adults bring to leadership. This first step is necessary to provide motivation for the difficult task of unlearning the reflexive behaviors that operate outside of consciousness to translate the old mindsets into action. Findings in neuroscience have thrown light on just how challenging this is. The old mindsets are congruent with mechanisms for perceiving and managing threats that have evolved to help guarantee survival (Hanson and Mendius 2009). But powerful leadership-development practices have evolved over the past 20 years to support such shifts in mindset (Kegan and Lahey 2009).

Skills

New mindsets are the foundation for new skills. Unless grounded in underlying mindsets, the skills will be used in service of the old paradigm, which will undermine not only their effectiveness, but also the credibility of the actor by sending mixed signals or even appearing to be hypocritical. Similarly, shifts in mindset can often have considerable impact even when they are not supported by new skills. Nevertheless, skills are necessary to fully leverage shifts in mindset and to increase the likelihood that people practice what they preach.

Making a comprehensive list of the needed skills is beyond the scope of this inquiry. However, we imagine it will be useful to many of our readers to see examples of the kinds of skills that have proved useful in supporting the cultural shift of which the Web is only one wave:

Self-Leadership: Leadership, like charity, begins at home. Just as it has become critical to understand systemic patterns in relationships, organizations and society, so too is it important to be aware of one's own internal system (Schwartz 1997). Long gone are the days when a person could "check her personality at the door" and act as if professional behavior is independent of personal character. A key element of self-leadership is emotional intelligence (Goleman 2006).

Interpersonal Skills: High-performance teamwork depends on high-quality communication. But habitual modes of speaking—whether polite or blunt—often obscure rather than enhance communication. Being able to understand the reality that others experience, and to enable them to understand your own, requires the two core skills of reflective conversation: advocacy and inquiry (Argyris and Schön 1974). Schein (2009) has recently offered the frame of "humble inquiry" as a core skill for those in the "helping" professions, and we see it as a core leadership skill as well.

Collaborative Leadership: As discussed throughout this report, a large set of forces are combining to give impetus to an inexorable shift away from one-way, hierarchical, organization-centric communication toward two-way, network-centric, participatory and collaborative leadership styles. According to Meister and Willyerd, by 2020, a "collaborative mindset" enabling "inclusive decision-making" and "genuine solicitation of feedback" will be not just advantageous, but also required (Meister and Willyerd 2010: 189). In many contexts, the primacy of individual intelligence will give way to the primacy of collective intelligence as leaders learn to take advantage of "crowdsourcing" (Howe 2009).

Network Leadership: As noted earlier in this chapter, there has been an evolution from exclusive attention to organization-centric leadership toward network-centric leadership. In our increasingly networked world, network-leadership skills will become as important as team-building skills. One enthusiast even argues that "the role of a leader ... is to make employees start to think in terms of their networks" (Anklam 2007: 226). Like teams, networks have predictable stages of development and other characteristics with which leaders need to be familiar (Hurley 2007: 20). However, leading networks is obviously quite different from leading teams of subordinates. As Boje (2001) has observed:

> Network leaders provide mediating energy. ... They set up exchanges between other partners, point out collective advantages in collaboration and identify dangers and opportunities. Leaders must be able to see and respond to trends and redirect energies as appropriate. They must be able to identify and bring together network resources to tie the network together and reconnect fractures.

Boje also adds that trying to exercise such leadership with the rules of a more traditional approach risks turning networks into "bureaucratic federations."

Small- and Large-Group Facilitation: An old but underutilized set of skills is that of group facilitation. The value of this skill at the small-group level has been well-documented, along with programs for developing it.[19] These skills are increasingly essential as leaders strive to elicit—and allow—leadership to emerge from teams and other groups. Being able to meet the challenge of managing virtual meetings, or even virtual teams, is more and more important (Lepsinger and DeRosa 2010).

Systemic "Hosting" Skills: Newer on the horizon are a variety of tools for convening at the system level. Although expert facilitators can always be engaged for high-stakes occasions, managers would do well to learn how to bring together stakeholder groups—and even a "strategic microcosm" of the whole system of those with a stake in an issue. As Peter Senge and his coauthors put it: "Extraordinary change requires building extraordinary relationships, and ... this requires gathering together diverse people representing diverse views so that they can speak and listen to one another in new ways" (Senge et al. 2010: 235). "Getting the system in the room," so to speak, even in this partial way, is a means of mobilizing a crucial mass of people willing and able to lead. This requires a repertoire of tools better suited to a network-centric, self-organizing world. Convening—or "the art of hosting"[20]—expands and transforms facilitation to create spaces in which generative conversations are possible and in which collective wisdom emerges. Practices that facilitate this kind of interaction include World Café (Brown and Isaacs 2005), Open Space Technology (Owen 2008) and Generative Dialogue. Effective convening also requires careful groundwork, and is likely to emerge only after "purposeful networking."[21] Likewise, it will also be aided by coming to understand the interests of the various stakeholders through "dialogue interviews" (Hassan and Bojer 2005: 19–22; Scharmer 2009: 241–243).

Systems Thinking: This is the "fifth" discipline that Peter Senge brought to the first of the meta-theories we briefly described above. It can be introduced conceptually by calling attention to some of the powerful archetypes that explain everyday patterns: the notion of unintended consequences, for example, or the "tragedy of the commons." Still, it is more powerful to teach this perspective experientially. Faculty members at M.I.T., the birthplace of systems thinking, have long used a simulation known as "The Beer Game" for this purpose. In our experience, Barry Oshry has developed more powerful exercises (Oshry 1999, 2007) by creating a highly powerful set of simulations and underlying frameworks based on a simple but effective model distinguishing between the worlds of "top," "bottoms," "middles" and customers. These simulations help participants see how their own reflexive reactions to the stresses of their place in a system lead them to think and act in ways that contribute to the counterproductive dynamics they experience but tend to blame on others. The classic version of the simulation is the "Organization Workshop," which consistently generates powerful insights into systems and one's personal experience of them. Substantial evidence supports its long-term impact.[22] Other simulations highlight the middle role, on which Oshry has done seminal thinking. In a remarkably intense and extended societal simulation, the "Power Lab," participants actually "live" in their roles for several days.[23]

Leading "Millennials": Organization-based leaders face a challenge in leading employees of the "Millennial" generation (born roughly in the last two decades of the 20th century) (cf., e.g., Shapira 2008). The experience of this generation in "growing up online" has led Gary Hamel (2009) to call it "Generation F—the Facebook Generation." He and many others note that members of this generation will expect the social environment of their work to reflect the social context of the Web (Tapscott 2008c; Meister and Willyerd 2010). As Hamel (2009) writes: "If your company hopes to attract the most creative and energetic members of Gen F, it will need to understand these Internet-derived expectations and then reinvent its management practices accordingly." He then provides a list of 12 "work-relevant characteristics of online life":[24]

- All ideas compete on an equal footing.
- Contribution counts for more than credentials.
- Hierarchies are natural, not prescribed.
- Leaders serve rather than preside.
- Tasks are chosen, not assigned.

- Groups are self-defining and self-organizing.
- Resources get attracted, not allocated.
- Power comes from sharing information, not hoarding it.
- Opinions compound, and decisions are peer-reviewed.
- Users can veto most policy decisions.
- Intrinsic rewards matter most.
- Hackers are heroes.

Effective leadership of Millennials must take into account these values and expectations, which constitute a cross-generational cultural difference. As David Weinberger has put it (Weinberger 2008: 70): "Attempts by traditional leaders to exhibit the traits of leadership will often strike (the) new generation as negative: an unrealistic attempt to control what is best left uncontrolled, a pathetic effort to retain one's power, a ridiculous attempt to puff oneself up." As we see it, the good news is that learning to lead Millennials is a good training ground for the leadership that is also suited to helping organizations adjust to the external environment in the face of accelerating change and unprecedented uncertainty.

Coaching: Coaching has been identified as one of the top tools for developing leadership. It will be even more important when it is operating in synchrony with the emerging leadership paradigm. On-the-job learning is the core of leadership development, and coaching helps ensure maximum value from such experiences. Developing the skills (and underlying mindsets) of coaching is also a good way for managers to make the transition from seeing themselves as critics to seeing themselves in the more supportive role of a coach. Instead of asking, "How could this person have performed better?" the coach asks, "How can I help this person learn from the experience?" For reasons such as these, Meister and Willyerd identify "developer of people" as one of five key areas of leadership that will be required in 2020 (Meister and Willyerd 2010: 189).

The belief that leaders are mostly made rather than born not only expands the notion of who can be leaders, but also the responsibilities of a leader. An important dimension of leading becomes the ability to cultivate the propensity for leadership in subordinates who have it naturally as well as supporting the development of those with a less innate talent for leading.

Knowledge

New areas of knowledge are also important for undergirding shifting mindsets and new skills. Two of the most salient are:

Web 2.0 Literacy: Emerging Web technologies are increasingly becoming a content area that leaders need to be informed about in order to be organizationally "literate."

To optimize their effectiveness, leaders will need to have at least a minimal degree of command over how to use these tools themselves. They will also need to know how to leverage these technologies to help other leaders grow. We believe that all people who wish to exercise leadership—regardless of their sector or their level in the hierarchy—would do well to follow the advice of the authors of *Megacommunities* (Gerencser et al. 2009: 205), who recommend that a community leader should "be familiar with the kinds of new media that exist. Such technologies provide ways of not only communicating within the megacommunity but of getting one's message out to a wide group of people via Web sites, wikis, written blogs, video blogs, texting and all types of multimedia programming."

To some extent, virtual communication will reduce the need for face-to-face interpersonal skills that require being able to act in the moment. (Given the unwillingness and/or inability of most people at all levels to master those skills, this is probably good news!) However, leaders will need to learn Internet-based interpersonal skills. The Web has enormous potential to expand both individual and organizational learning. It will be increasingly easy to listen simply by engaging in the skillful use of Web 2.0 tools. Social media, RSS feeds and other tools will likely expose leaders to a broader range of information than has been previously possible, from perspectives other than their own. However, mere exposure to other perspectives does not ensure effective engagement. To constructively manage encounters with perspectives based on differing assumptions about the world, leaders will need even more skill in "listening" to other views, constructively asserting their own and being willing to challenge their own assumptions. The inability to do this may result in a marked decrease in the volume and quality of information voluntarily made available to them. Likewise, without an openness to learning, leaders risk using the new media to seek information that merely confirms their biases.

Web skills also include "netiquette"—that is, Internet etiquette. Examples are being judicious about when to hit "reply" versus "reply all" and avoiding the use of "all caps," which has come to be perceived as aggressive.

Cultural Literacy: The Web increases the need for leaders to be sensitive and able to manage differences in national as well as organizational cultures. The Web will increasingly allow teams and networks to be virtual and to include individuals from different countries, races and religions. Effective leadership will need to take those differences into account. Anyone wishing to influence a cross-cultural team will need to be aware of the need to create "cultural islands" in which people can construct at least a minimal foundation of mutual understanding and trust (Schein 2010).

Fostering New Cultures

Schein (2010) has observed that there is no single "best" culture. While this true, the case is getting stronger and stronger for a culture in most organizations in which openness, transparency and collaboration are among the core values. Such a culture is well suited to learning and innovation, is often attractive to employees (depending, of course, on the national and organizational culture), is motivating (again, depending on culture), and offers well-documented benefits (Li 2010).

Yet we learn from Schein that culture change is very hard, that it has in most cases evolved rather than being chosen, and that it is remarkably hard to "create." He defines culture as "a pattern of shared basic assumptions learned by a group as it solved its problems of external adaptation and internal integration, which has worked well enough to be considered valid and, therefore, to be taught to new members as the correct way to perceive, think and feel in relation to those problems." (Schein 2010: 19). Cultures change only when an organization learns a new way to solve its particular challenges regarding adaptation and integration.

Whether and how an organization should aspire to radical cultural change is a question we address in Chapter 4. However, at a minimum, anyone who wishes to exercise leadership effectively in the emerging social and organizational worlds must understand the cultural trends likely to shape the future. Furthermore, most leaders would do well to embrace the elements of this culture to at least some degree. To be sure, many organizations are large and complex enough to have more than one culture, each suited to the particular challenges and history of its function. At the same time, even complex and geographically diverse organizations often have a distinctive culture. Thus, it is clear that culture matters. As one wise observer has said: "The soft stuff is the hard stuff."

In our experience, there are at least three paths to addressing the gap between an existing and a desired culture:

- *Gradual/evolutionary:* people discover they can succeed in new ways (Schein 2010);
- *Gradual/directed:* policies and directives change mindsets and behavior over time;[25]
- *Accelerated/orchestrated:* intensive, top-down driven, transformative initiatives.[26]

Whether one of these approaches, a hybrid or something entirely unique makes sense in a particular case depends on many factors. In Chapter 4, we offer guidelines on how to approach this question.

3 The Organizational Impact of the Web by Sector

How the Web is Impacting the Business Sector

Challenges and Opportunities

The most extensive applications of Internet-based technology are in the business sector. A study in early 2010 found that 72 percent of Fortune 500 companies were using some form of such tools.[27] Moreover, a study from a few months earlier found that 95 out of 100 "top brands" were making use of Web 2.0.[28] By contrast, studies of Web 2.0 usage in government (Eggers 2007) and in the social sector (Kanter and Fine 2010) suggest that organizations in those sectors are lagging behind.

Thus, the impact of the cultural changes enabled by Web technologies is gaining momentum in the world of business and transforming traditional patterns of interaction and communication in multiple domains. A wide range of experts and practitioners see dramatically shifting patterns in many areas, including the following:
- Between employees and management within companies;
- Among employees within companies;
- Between companies and talent outside the company;
- Between companies and their customers and suppliers;
- Between companies and their competitors;
- Between companies and organizations in other sectors;
- Between a "new industrial era" of networked, virtual companies and their customers.

Enterprise 2.0: The Open, Networked Enterprise

A new terminology is evolving for describing the resulting dynamics within companies and with their customers in such an environment. The term "Enterprise 2.0" has been widely adopted (McAfee 2006a). In an essay of that name, Don Tapscott[29] suggests that these patterns point to a new paradigm, the "Open, Networked Enterprise" (Tapscott 2008a). This and a growing body of related work highlight radical shifts from the traditional model of a corporation in many areas, including the following dozen:

- *Innovation:* from closed and done within the company to open and including both co-creation with customers and receptivity to ideas from a global brain trust;
- *Intellectual property:* from proprietary and protected to open and shared;
- *Knowledge:* from "stocks" (e.g., books and libraries) to "flows" (e.g., conversations in which bits of contextually relevant information and tacit knowledge are exchanged);
- *Information management:* from opaque and asynchronous to transparent and in real time;
- *Corporate boundaries:* from closed and vertically integrated to open and networked;
- *Bottom-line measurement:*[30] from ROI ("return on investment") to ROR ("return on relationships")[31] and ROC ("return on collaboration");[32]
- *Marketing:* from one-way "push" strategies to two-way conversations;[33]
- *Planning:* from visioning and strategic planning to sensing and adapting;[34]
- *Strategy:* from "make and sell" organizations to ones that "sense and respond";[35]
- *IT capacity:* from company-based to the resource pool of "the Cloud";[36]
- *Internet access:* from PCs and search engines to mobile phones and apps;[37] and
- *Internal authority distribution:* from hierarchical and top-down to shared influence or even inverted power relationships.[38]

Of course, the pace of change varies depending on many factors. For example, large companies find it difficult to move quickly to the cloud owing to their considerable capital investment in legacy technology.[39] This is probably even truer outside the United States. Indeed, Elmar Hussmann, an innovation expert at IBM Germany told us: "When speaking to customers, there is still a lot of reservation toward the cloud because of concerns about data protection." He reported that the degree of concern varies somewhat by country and national regulatory environment, with companies in Germany being "extremely concerned" and taking a "particularly conservative approach." By contrast, he added that there are "already startups that base their entire infrastructure on the Cloud."[40] Analyzing a survey specific to Germany, Nicole Dufft concludes her analysis of a survey of companies conducted in the summer of 2007 with the following observation (Buhse and Stamer 2008: 141):

> When we look at major, knowledge-intensive companies and see that a mere 2 to 6 percent use Web 2.0 tools in a company-wide fashion, then we can hardly describe these tools as "a normal part of our business environment." Moreover, without a clear perception of the usefulness of Web 2.0, a major prerequisite is missing for the spread of Enterprise 2.0 ideas within German business.

Nevertheless, data on the usefulness of Web 2.0 tools is becoming harder and harder to ignore. They are not only enhancing the effectiveness of individual companies, but changing fundamental patterns of cooperation and competition in ways that are bene-fiting companies and consumers alike—and the pace of adoption seems only likely to increase. For instance, regarding the critical example of the cloud, the lead article in the Oct. 23, 2010 issue of *Barron's* was "Big Companies Adopting Cloud Computing Quicker than Predicted," and it cited the emerging option of "private clouds" as a less-risky transitional step.

Ecosystems of competition and cooperation

The cumulative consequence of these changing patterns is that the Web is the driving force in the creation of ecosystems eroding the traditional boundaries around compa-nies and fostering new forms of virtually clustered and networked organizations. In the new environment, successful enterprises offer customers not just a product or service, but a *platform* capability upon which they can build their own value proposi-tions. Amazon.com and eBay are good examples of such platforms. They provide a fundamental capability (e-commerce) that allows thousands of merchants to set up shop more quickly and with more innovative value propositions than they could other-wise provide by themselves. The most visibly successful companies that have created such ecosystems are in the digital realm (e.g., Google, SAP and Facebook, in addition to the two just mentioned). But they signal a trend that is unstoppable (Tapscott and Williams 2006, 2010).

In these ecosystems, relationships are a two-way street. Companies work with their suppliers to directly add operational advantage through win-win processes. They don't just define suppliers as services, but also define their own operations as services to the suppliers. They reject the Darwinian model whereby only one supplier can win out over others (and its "reward" is to have its margins squeezed mercilessly by cus-tomers), and replace this with a collaborative model built on services. Information shared in this way then becomes the foundation for continual process improvement, newly discovered market opportunities and new response mechanisms (Mullholland, Thomas and Kurchina 2007).

In such ecosystems, companies mimic the biological example of "keystone spe-cies" that proactively maintain the health of the entire ecosystem for the ultimately self-serving reason that their own survival depends on it (Iansiti and Levien 2004). They view their suppliers as channels to new markets via their own ecosystems. Com-panies that have adopted this approach have a powerful advantage over those that haven't, as illustrated by Wal-Mart and its relentless demands on its partners for ever-greater supply chain integration (Mulholland, Thomas and Kurchina 2007: 62).

In these ecosystems, companies use Web tools to "mash up" organizational struc-tures, creating dynamic new sources of business and radical economies in existing business relationships. In this way, the Web is stimulating the creation of next-gener-ation business models that are radically reshaping the competitive environment. In-

stead of having businesses compete with one another individually, networks of businesses are now also competing with one another. This is a win-win arrangement that
creates lower costs for consumers while creating new markets and diversifying risk
for individual companies.

Such ecosystems also erode the boundaries between producers and consumers.
Customers do more than customize or personalize their wares; they can self-organize
to create their own. The most advanced users no longer wait for an invitation, but instead become "prosumers," a term (coined by Tapscott and Williams 2006) for the
blend of producer and consumer who shares product-related information, collaborates
on customized projects, engages in commerce, and exchanges tips, tools and "product
hacks."

In the emerging environment, companies are increasingly forced to operate on the
"edge of chaos." As they adapt to and experiment with the radical new opportunities
created by the Web, it is increasingly hard to demarcate traditional boundaries between themselves and their environments. This trend, which has been noted by scholars, was confirmed by our interviews with managers in the United States and Germany. Elmar Hussmann, the IBM expert, told us: "In the IBM research and
development organization, it is increasingly difficult to draw a clear separation line
between inside and outside. We are engaged not just with our own internal networks,
but also in alliances, doing joint development, for example, on open standards and on
policy activities. It's a 'blurring borderline.'"[41]

The New Industrial Revolution

A further factor driving companies into the relative safety of ecosystems is the arrival
of what a recent issue of *Wired* heralded as the "New Industrial Revolution" (Kelly
2010). This term refers to the ways in which the Web has enabled radically expanding
opportunities for entrepreneurs to compete with established businesses by taking advantage of technological innovations such as "3-D printing" in combination with
Web-enabled virtual networks. As Ashlee Vance (2010) put it in the *New York Times:*
"A wealth of design software programs, from free applications to the more sophisticated offerings of companies including *Alibre* and *Autodesk*, allows a person to concoct
a product at home, then send the design to a company like *Shapeways*, which will
print it and then mail it back." This is good news for many players, including customers (who get better service and lower prices), entrepreneurs (who have an expanded
menu of options) and even society as a whole (through greater productivity). Nevertheless, for companies wedded to traditional ways of doing things, this is bad news,
because they face new threats that are likely to reduce market share and profits, and
could very well even put them out of business.

The dilemma of intellectual property

In the struggle to remain competitive, one of the chief dilemmas companies face is how to handle the Web's tendency to create unfettered—and increasingly "free"—access to information and intellectual property, as reflected in the popular (and misunderstood) slogan "information wants to be free."[42] David Weinberger, one of the authors of *The Cluetrain Manifesto* (2000), the book that heralded the game-changing nature of Web 2.0, has since written (Buhse and Stamer 2008: 68):

> Traditional companies have assumed that they can control their customers by selectively releasing information … (but) the Web has put customers in touch with other customers. In the new environment, attempts to re-exert control over markets can have quite a negative effect. … On the other hand, if a company is willing to give up on trying to keep the market conversations relentlessly positive, if they are willing to enter into those conversations as people who acknowledge their vested interests, but who have some genuine passion for their products, they can benefit greatly from the growth of networked markets.

In this new environment, commercial companies are potentially disadvantaged when it comes to collaborating, because they are hampered in sharing knowledge if they hold onto traditional notions of maintaining intellectual property (IP) rights. Historically, the default path for a commercial company has been protection of intellectual property, including the use of secrecy.[43] The new challenge is to reap the benefits of collaboration and open source methods while redefining the boundaries of proprietary IP.[44] Unfortunately, as one expert has noted, in Europe (and surely elsewhere), "there is protection of work for 90 years, in a society that develops faster and faster!"[45]

Despite the caution that these concerns provoke, a number of companies are learning to share some—though by no means all—of their IP for reasons having to do with the bottom line: to increase demand, foster relationships and stimulate progress in other areas. According to Web expert Anthony Williams, companies like *Novartis*, *BMW*, *IBM* and *LEGO* "are becoming a new kind of business entity, one that co-innovates with everyone, especially customers, shares resources that were previously closely guarded, harnesses the power of mass collaboration, and behaves not as a multinational, but as something new: a truly global firm" (Williams 2010a: 5). In his view, which is increasingly widely shared, "no single company, whatever the industry, can create all the innovations needed to compete. Individuals and companies are deploying new knowledge in unpredictable ways. To harness this innovation you need a lot of partners, and a lot of people developing designs and putting them together as customer solutions. This means tapping into a broad ecosystem, and it means opening up some of your IP" (Williams 2005: 1). Williams reiterates the concept of an "Open Networked Enterprise," which was promoted elsewhere by his coauthor, Don Tapscott, in 2006 (cf. Buhse and Stamer 2008). He summarizes his view in the following way (Williams 2010a): "Notions of intellectual property are changing and will change even more. Clever companies will manage their intellectual property like mutual funds—with some IP highly protected and other IP shared with the world free of charge."

In the next section, we provide illustrations of how companies among the so-called early adopters are providing evidence that promises eventually to lure skeptics onto the Web 2.0 bandwagon.

Patterns and Examples

Below, we provide a number of examples of how the Web is used in the business sector, organized according to some of the most salient patterns of application.

Communication Between Employees and Management

Enhancing communication- and information-sharing from the top

- Corporate blogging has now become a competitive sport, with regular rankings of the top corporate blogs. In 2006, there were 124 corporate blogs worldwide. In 2010, it would be hard to find a major corporation without one, and many CEOs have personal blogs.
- *Alcatel-Lucent*, in Germany, has created an internal *YouTube*, on which the most-watched video shows an employee explaining the firm's strategy. The chairman of the board has said: "We are learning from the 'Generation Internet' how to communicate, inform ourselves and collaborate in a 'totally networked' way."[46]

Enhancing communication from the bottom up

- At *NEC Renesas*, a Japanese semiconductor company with 47,000 employees, a CEO used a blog to ask employees what the strengths of the firm were. Doing so led to an active discussion that exposed the CEO to new perspectives.[47]
- Through *Toronto General Hospital's* "Rypple," teams receive anonymous feedback in response to one question a week (e.g., "What one thing can we do to reduce readmission rates?"). The director of innovation reported that "the process opened up the feedback channels among the different hierarchies, and we were able to set up a model (with which) we could create continuous team improvement" (Li 2010: 67–68).
- *Communispace* manages a virtual community of 400 employees for a major financial services company, enabling feedback on initiatives regarding critical strategic decisions (ibid.: 55).

Moving decision-making away from the top and center

- Through "distributed decision-making," companies are pushing decisions away from the center and closer to the customer, where relevant information and knowledge resides. The defining paradigm for this model is *Mozilla*, which created the *Firefox* browser. The principal task of Mozilla's 170 employees is to coordinate the contributions of thousands of people who (at no charge) build and market Firefox (ibid.: 40). Volunteers supply 50–60 percent of all patches. Nevertheless, the process for decision-making is highly prescribed. The work is divided into 100 "modules" led by "module owners," who are the only ones who can authorize changes to the code and are oftentimes not even Mozilla employees.

Reducing the barriers to self-organization

- Although predicted years ago as a pattern within business (Wheatley 1992), until only recently, self-organizing was seen as a fringe phenomenon peculiar to communities that seek to download MP3s or to highly specialized communities such as software development, where the open source movement has been a poster child for self-organization through its widespread, distributed software development practices (Tapscott 2008a: 108). However, technology has dramatically lowered the costs of self-organization.
- *EBay* illustrates how a top-down structure enables bottom-up self-organization. In addition, companies as diverse as *BM, Eli Lilly, Harley-Davidson* and *Procter & Gamble* "are discovering how to make self-organization a key component of the modern day strategic arsenal" (Tapscott 2008a: 32, citing Ticoll and Hood 2005). It succeeds because it "leverages a style of peer production that works more effectively than hierarchical management or open markets for certain tasks." However, the impact is most visible in the production of information-intensive goods and services, as seen in software production, media, entertainment and culture (ibid.: 108).

Communication and Collaboration Among Employees

Employee peer-to-peer personal and social communication

- Social-networking websites in companies have become a popular way to interact among employees. Likewise, there is widespread use of instant messaging and sites like *Facebook* by employees who are geographically separated from their co-workers and want to keep in touch.[48]
- *IBM* offered a free internal service to employees for blogging, resulting in over 1,000 active blogs, and the *BBC* reports 600 active bloggers among its employees (Tapscott 2008a: 103).

Sharing knowledge and best practices among employees

- *BT* (formerly British Telecom) created "BTpedia,"[49] a companywide wiki tool designed to democratize the publication process and elicit informal knowledge. It also features internal *YouTube*-style podcasting that allows employees to upload short video or audio "learning nuggets."[50]
- *Best Buy's* "Geek Squad" used an employee-initiated informal group playing the online multiplayer game "Battlefield 2" as a vehicle for sharing technical best practices (Tapscott and Williams 2006: 241–242).
- In 2009, *Yum! Brands*, the world's largest restaurant company and the parent of *A&W Restaurants, KFC, Long John Silver's, Pizza Hut* and *Taco Bell*, created a community platform called *iChing* using *Jive Software*. Using this platform, 6,000 corporate restaurant employees around the world are able to pose a question at the end of their shift, and they typically find a response by the next morning (Li 2010: 30).
- *Premier Farnell plc*, a U.K.-based multinational marketer and distributor of electronic products to engineers, distributed several thousand video cameras among 4,100 employees, encouraging them to record their best practices and upload their videos to an internal site, called "OurTube." Employee contributions were not screened, which has reportedly brought about a profound change in the company culture (ibid.: 32).
- *Synaxon AG*, in Germany, has made all of its work guidelines and regulations accessible via a wiki system (Buhse and Stamer 2008: 64)
- *Transunion*, a large credit-reporting company, used *Socialtext* to try to keep employees from sharing information on *Facebook*. However, once the system was set up, employees used it primarily to ask each other questions, and the questions and answers were recorded in a database. Other tools allow people to vote on their favorite answers, analyze the answers chosen to solve problems and also analyze which answers correlated with the issues most vital to the company (Li 2010: 93).

Enhancing internal collaboration

- *Transunion* saved $2.5 million in deferred IT spending in less than 5 months at a cost of $50,000, which went toward installing the enabling social software (ibid.: 93).
- *IBM* created a Technology Adoption Program (TAP)[51] to foster "emergent, community-oriented collaboration patterns" (IBM 2009). It enables an intra-enterprise crowdsourcing approach accessible to the company's 350,000 employees worldwide. A report on TAP says: "Like many large companies, IBM is organizationally siloed. The TAP helped to flatten out IBM by triggering the development of Early Adopter communities that bridged those silos" (ibid.: 7).
- The technology company *SunGard* used *Yammer*, an internal version of *Twitter*, to enable employees to write each other brief messages. This has led to sharing project

information that has in turn accelerated product development. Anyone in the company is able to start a Yammer network and invite colleagues without corporate sponsorship or permission. In fact, it proved "so effective in speeding up product development that SunGard rolled out Yammer to 20,000 employees across thirty countries, where it is beginning to affect all aspects of operations from sales to customer service" (Li 2010: 28). This is a good example of an "emergent" phenomenon.

Cutting costs through collaboration

- *Cisco Systems* reported large cost savings from collaborative initiatives. Most came from reduced travel through use of in-house video-conferencing tools. Less travel also resulted in greater productivity and faster results (see the case study on Cisco at the end of this section for a more detailed estimate of ROI through collaboration and other Web-based initiatives) (ibid.: 93).
- *SolarWinds*, a network management software provider, built a 25,000-member user community of network administrators who help each other with problems large and small. This enables them to support a customer base of 88,000 companies with just two customer-support agents (ibid.: 29).
- The drive to lower costs has spurred organizations to rely on Web tools such as blogs and social networks (and the site *Gumtree*) for recruiting instead of headhunting services (Bodie 2009).

Leveraging the efficiencies of the cloud

- *SmugMug* is an online photo-sharing service with 15 employees, 150,000 customers and 72 million photos. It is one of an increasing number of companies that take advantage of Amazon.com's Simple Storage Service (S3). In a period of just three months, SmugMug saved $500,000 by using S3 instead of adding new servers (Mulholland, Thomas and Kurchina 2007: 48–49).

Reducing the cost of failure

- Open source projects, such as the computer software system *Linux*, lower the cost of failure. Shirky concludes from this and similar examples that "services that tolerate failure as a normal case create a kind of value that is simply unreachable by institutions that try to ensure the success of most of their efforts" (Shirky 2008: 248).
- *Meetup*, an online service for organizing meetings, "has been consistently able to find new offerings without needing to predict their existence in advance and without having to bear the cost of experimentation" (ibid: 2008: 248).[52]

Access to Talent Outside the Company

Extension of organizational boundaries through crowdsourcing[53]

- "The mining firm *Goldcorp* made its proprietary data about a mining site in Ontario public, then challenged outsiders to tell them where to dig next, offering prize money. The participants in the contest suggested more than a hundred possible sites to explore, many of which had not been mined by Goldcorp and many of which yielded new gold. Harnessing the participation of many outsiders was a better way ... than relying on internal experts" (Shirky 2008: 247–248, citing Tapscott and Williams 2006).[54]
- *Netflix* conducted an open competition with a reward of $1,000,000 for the best collaborative filtering algorithm to predict user ratings for films based on previous ratings.
- A U.S. biotech company that was unable to find a solution for a new DNA-sequencing test method posted the problem on *InnoCentive*. The company found a high-quality solution within four weeks from a Finnish research team in an entirely different field (Monitor Institute and David and Lucile Packard Foundation 2007: 10). *Alcatel-Lucent, Siemens* and *Unilever* have all used *InnoCentive* as a broker for crowdsourcing solutions to R&D challenges (Williams 2010a: 5). The company provides access to 160,000 problem solvers (including retired, unemployed or aspiring scientists) in 175 countries. Foundations have also turned to InnoCentive, and with good results (see the examples under "Crowdsourcing"). Organizations that provide a similar service include *NineSigma, Elance* and *YourEncore* (Tapscott and Goodwin 2008: 9).
- The German startup *Atizo* administers and markets a Web community of creative thinkers with specialized expertise. *Atizo* develops innovation-management tools to support this community and other teams of innovators in organizations of all sizes and across all sectors.[55]
- In the world of journalism, the *New Assignment* project was launched to demonstrate that "open collaboration over the Internet among reporters, editors and large groups of users can produce high-quality work that serves the public interest, holds up under scrutiny and builds trust." It resulted in the publication of seven original essays and 80 interviews as well as a series of stories about collaborative journalism for *Wired* magazine (Noveck 2009: 2).
- *Procter & Gamble* moved beyond its time-honored principle of "grow from within" to address the problem that, in 2000, only 15 percent of its products were successful. The CEO created a new program called *Connect + Develop* (i.e., connect externally to find new ideas, then develop them internally), which aimed to see 50 percent of the company's new products developed externally. It also launched a new website, pgconnectdevelop.com, which highlighted research needs the company wanted to address and encouraged contributions from the approximately 2 million

researchers working on related issues worldwide. These days, 65 percent of its new products succeed, and 35 percent are sourced externally—with no increase in R&D costs.[56]

Communication and Collaboration Between Companies and Customers

Leveraging "prosumers" (consumers who are also producers[57])

- *CNN* maintains *iReport.com* as an online site with user-generated content. Anyone can upload a video, and CNN staff members sift through them to select a few to feature (Li 2010: 31).
- *Second Life* members participate in the design, creation and production of the product, a virtual environment that serves as a "massively multiplayer online game" (MMOG) (Tapscott and Williams 2006: 125–126). *Linden Labs*, its originator, produces less than 1 percent of its content, getting up to 23,000 hours of free development from users every day.
- As a novel approach to developing GPS features for future cars, *BMW* released a digital design kit on its website to encourage interested customers to design them. Thousands responded with ideas, many of which have been incorporated. As a result, *BMW* now hosts a "virtual innovation agency" on its website, where small and medium-sized businesses can submit ideas in the hope of establishing an ongoing relationship with the company (ibid.: 128–129).
- A vibrant "prosumers" community has formed around products of *LEGO*, which is increasingly focused on high-tech toys (e.g., robots). The products appeal not just to teenagers, but also to hobbyists who enjoy improving them (ibid.: 130). The company uses "mindstorms.lego.com" to offer a free, downloadable software-development kit to customers to encourage tinkering. Each time a customer posts a new application, the toy becomes more valuable.

Transforming customer relations management (CRM)[58]

- The evolution of *Flickr*—one of the original Web 2.0 services—is a fascinating example of how social media can shape not just the effectiveness of an organization, but also its very purpose. Flickr's founders, operating under the name Ludicorp, had begun their enterprise as a site for gaming. Users "told the company to shift their business focus from gaming to online photo-sharing, and the company listened" (Shuen 2008: 7).
- *Comcast* is using *Twitter* as a customer-service tool. A "director of digital care" oversees a Twitter account (@comcastcares), which solicits feedback and concerns.

Marketing

- Using *Twitter*, *Dell* posted Twitter-only deals to its twitter.com/delloutlets page early in 2009. The number of people following Dell mushroomed to 600,000 within 6 months and to 1.6 million by the end of the year. Sales increased substantially, generating a "tremendous" ROI (Li 2010: 84). Visitors to *Dell Outlet* are engaged in consultative dialogue. Many other companies have since followed Dell's lead.

- To promote its Fiesta model, the *Ford Motor Company* gave a free car to 100 people to drive for six months, based on an online competition involving points for each video, blog post, tweet or photo uploaded to a company website, and for each comment on sites like *YouTube*. The results were "astounding" by a number of hard measures.[59] Based on successes like this, Ford has shifted a quarter of its marketing spending to digital and social media owing to their ability to "deeply engage people in a way not previously possible" (ibid.: 85).

- *Nokia* has increased the effectiveness of its marketing while reducing costs by shifting from advertising in traditional media to more modest investments in social media, as well as by the parallel shift from "bought media" (the purchase of promotion) to "earned media" (direct communication and conversation with customers). This is an example of what *nGenera* researcher Dennis Hancock calls the "incredibly shrinking marcom (marketing communication) expense line." The shrinking investment required for comparable impact is possible because social media enable "ambient intimacy." People can maintain more "weak-tie" network relationships than ever before. Many customers are opting to forge such relationships with various brands. Through these linkages, brand-related information can spread through so-called social graphs in a variety of ways. "It's like word-of-mouth marketing on steroids," Hancock says. Moreover, compared to traditional marketing methods, the associated costs can be trivial.[60]

Proactive monitoring of customer perceptions

- The importance of tracking customer perceptions and having the ability to quickly respond is illustrated by the dramatic (albeit rare) cases in which negative information about a company spread via social media has actually been able to negatively impact financial performance. This has happened at *Dell* ("Dell Hell"), *United* ("United Breaks Guitars") and *Radian6* (Li 2010: 14). Negative feedback of the kind illustrated by these unusually visible examples is what propelled many companies to recognize the importance of joining the social Web (Morgan 2010a).

- Companies can monitor what customers are saying about their organization through free tools such as *Google Blog Search* or *Twitter Search*. For a fee, they can get real-time monitoring through a number of vendors such as *BuzzMetrics*, *Cym-*

fony, Radian6, Umbria and *Visible Technologies* (Li 2010: 54, citing open-leadership. com as a further source).

- Companies can also supplement focus groups through online communities. Vendors such as *Communispace, Networked Insights* and *Passenger* pull together groups of up to several thousand people for idea generation, feedback, live chats and other forms of interaction (ibid.: 55).

Engaging customers as co-designers of products and services ("prosumers")[61]

- *Dell* created "IdeaStorm" to pinpoint the problem behind the company's languishing sales. Customers post ideas, build and vote on other contributors' suggestions, and talk with each other and *Dell* representatives. *Dell* is transparent about how it is integrating customer feedback into new products and services. Tapscott and Goodwin conclude from this example that (Tapscott and Goodwin 2008: 7): "Listening to and collaborating with customers is good; acting on their advice is better; and showing them exactly how you are paying attention is best. ... New technology makes it easy to do the first part, but strategic and cultural shifts are required to do the second and third."

Responding to crises in public/customer relations

- *Best Buy's* "Spy"[62] allows it to monitor mentions of the company on the Web. Chief Marketing Officer Barry Judge tracked it in real-time on a large TV in his office (as did CEO Brian Dunn). This enabled Judge to swiftly intervene to diffuse a firestorm of customer protest resulting from the mistaken mailing in September 2008 of an exclusive offer to all members rather than to the intended target audience made up of its best customers (Li 2010: 235–236). This contrasts sharply with the inept responses of *Eurostar* (regarding customer frustration over train delays), *Virgin Atlantic* (which had a "public relations nightmare" when some cabin crew members posted derogatory comments about both the airline and passengers in a Facebook forum) and *British Airways* (which had a similar problem with negative Facebook postings about passengers) (Meister and Willyerd 2010: 146).

Communication Between Companies and Suppliers

- Dell gained a 32 percent market share in the PC business by leveraging its relationships with suppliers. Such success derives from its ability to assemble a batch of components within hours and to undersell its competitors. Dell achieves this by assessing its inventories on any given day and then asking members of its supplier network what they can provide and at what price within the next day. Dell then offers discounts to its customers on the components it purchased from the lowest bidder (Mulholland, Thomas and Kurchina 2007: 66).

Communication and Collaboration Among Companies

Creating platforms on which other companies can do business

- *Amazon.com* has created *Simple Storage Service* (S3), an extension of its massive data-storage infrastructure. It charges organizations wholesale rates for access to simple, cheap and infinitely scalable storage. Its customers include startups such as *ElephantDrive, MediaSilo, Plum* and *SmugMug* (ibid.: 48)

Exporting innovation through online marketplaces

- In 1999, the online technology transfer marketplace *Yet2.com* emerged to enable companies to make available for sale or license intellectual property that had been developed internally but was not being utilized (Tapscott and Williams 2006: 103).
- "Oracle saves $550 million annually by letting customers serve themselves, estimating that each customer-care call handled by an employee costs $350, compared to about $20 for those done on its website. Similarly, by moving 35 million customer service calls to a Web-based self-service environment, IBM saved hundreds of millions of dollars a year" (Eggers 2007: 30).

Ecosystems of Partners, Suppliers and Customers

- *Salesforce.com* was one of the first SaaS (software as a service) companies. CEO Marc Benioff extended his platform to make it a broader ecosystem by introducing the *AppExchange* platform, with the result that over 250 unique SaaS solutions are now available and more than 50,000 objects and applications have been created by salesforce.com customers (Mulholland, Thomas and Kurchina 2007: 65). In a related example, *SAP* created "EcoHub," making all its solutions available and enabling "partners in its ecosystem, such as systems integrators or technology or software partners, to provide additional background or solutions of their own" (Li 2010: 66).

- *Lastminute.com*, which specializes in the sale of plane, hotel and other tourism-related commodities about to expire, has aggregated 13,600 suppliers (e.g., airlines, hotels, etc.) that pass on information about expiring seats or rooms to lastminute.com for packaging and selling (Mulholland, Thomas and Kurchina 2007: 66).

Creating new business models

- In just 10 years, the Mumbai-based *ICICI Bank* has become India's second-largest retail bank, leading in every retail product market that it targets. ICICI drives over 70 percent of its transaction volume through electronic channels in a country where Internet and mobile-phone penetration rates are below 5 percent. It achieves this via e-lobbies and unstaffed branches where customers help themselves. Having IT systems that are free from legacy issues enabled the bank's investments to be less than 10 percent of developed country benchmarks (Williams and Goodwin 2008: 6).

Using B2B peer production

- As Williams and Goodwin write (ibid.: 8): "A new breed of value-chain partner is, at once, consumer and producer. Companies in nearly every sector of the economy, from software (*Linux* and open source) to consumer products (*Procter & Gamble*) and manufacturing (the Chinese motorcycle industry), are embracing new models of collaborative innovation." The principles of "peer production" are used to engage and collaborate in a non-hierarchical, self-organizing manner with a focus on improved customer service, brand loyalty and enhanced innovation. Rather than relying solely on in-house resources to develop new products and services, leading enterprises are harnessing external ideas, resources and capabilities. Producers in a wide variety of industries are responsible only for final product assembly and marketing. They rely on peer production to tap up to hundreds of firms to help design and build finished products. Overall, this collaborative approach enables risk-sharing and allows the network to tap diverse skills and resources.
- Expanding on the pattern in the preceding example, the Chinese motorcycle industry now dominates the Asian market, making half the world's motorcycles. Although this is partly due to state-run operations, many of the most impressive innovations are coming from private-sector upstarts *Zongshen* and *Longxin*, which use mass collaboration for competitive advantage. They rely on modularity, a highly iterative process and localized manufacturing concentration. High-level designs are set out in rough blueprints that enable collaborative suppliers to make

changes to components without modifying the overall architecture. Manufacturers try out new designs in rapid succession with suppliers rather than the assembly company, and they assume responsibility for ensuring component compatibility in design and manufacturing (Tapscott and Goodwin: 2008: 8).

• *Novartis, BMW, IBM, LEGO* and many others are examples of a new kind of business entity—one that "co-innovates with everyone (especially customers), shares resources that were previously closely guarded, harnesses the power of mass collaboration, and behaves not as a multinational, but as something new: a truly global firm" (Williams 2010a: 5).

Using mashups for business intelligence (BI)

• Multinational pharmaceutical firm *Pfizer* uses an intranet BI mashup for product management support. The mashup supports ad hoc querying, forecasting, planning and modeling for executives making resource-investment decisions (Kobielus 2009: 12).

• Spanish financial services institution *Caixa Galicia* applies Internet mashups for mortgage brokerage support. The mashup provides mortgage-brokering sales staff with access to data from both internal (e.g., customer mortgage records) and external (e.g., housing prices and conditions in various markets) sources (ibid.: 12).

Creating and tapping previously unprofitable niche markets in "the long tail"

• The cost advantages of the Internet make it possible to profitably sell products for which there is not a large market. Tapping previously unprofitable niche markets in this way was dubbed "the long tail" by Chris Anderson (2008), who cites *Amazon.com* and *Netflix* as examples of businesses applying this strategy. A product like *iTunes*, with a digital catalogue and digital product, is able to get "all the way down the tail" by leveraging the near-zero marginal costs of manufacturing and distribution.

Creating cross-sector ecosystems

• Although *Stonyfield Farms* is a for-profit enterprise that is the leading producer of yoghurt in the United States, it defies the logic of traditional capitalism by choosing to pay the farmers who supply milk 100 percent more than the going market rate. That's possible because of Stonyfield's commitment to creating a healthy and sustainable life and livelihood for all members of its ecosystem. It wants the farmers who supply organic milk from cows that live under "humane" conditions to thrive. It remains competitive while doing this by relying on word-of-mouth marketing and saving the money that would have otherwise gone into advertising (Gunther 2008).

Cases in the Business Sector

The greatest penetration of the culture and tools of Web 2.0 is to be found in the United States. Thus, it is not surprising that the corporate example that many observers single out as representing "best practice" in this domain, Cisco Systems, is in that country. For this reason, we think it makes sense to lead with a profile of that example. However, because we see our primary audience for this report in Europe—and, in particular, Germany—we have identified a cluster of cases from that country as well. Finally, we include a case that illustrates how superficial use of the Web can be of no help in a crisis.

Comprehensive Best Practice Example

- *Cisco Systems,*[63] a high-tech company based in California's Silicon Valley, provides an illustration of how one company has deeply embraced Web technology to transform the way it does business, creating a platform that supports an ecosystem of other stakeholders. According to the official record, the company is making an explicit attempt to evolve a "next generation company—Cisco 3.0, reinventing itself around Web 2.0 and then taking the lessons learned to its customers." The internal foundation of this new company is a set of significant changes in organizational structure that distribute decision-making, innovate faster, bring products to market sooner and capitalize on market transitions. After the 2001 technology bubble burst, Cisco CEO John Chambers created a system of councils and boards that enabled Cisco to shift decision-making down several levels. Nine councils of about 16 executives each report to the top team. Working groups report to 50 boards, which, in turn, report to the councils. About 750 executives are currently involved in the councils and boards, participating in such strategic decisions as acquisitions, entry into new markets and creation of new products. Leadership within the councils is typically shared between two people (e.g., one from sales and another from product development or engineering).

Structured for participation

Consultant Charlene Li reports what she describes as "shocking" numbers regarding what the company accomplishes through this structure for participation (Li 2010: 43). The CEO says Cisco operates better "as a distributed idea engine where leadership emerges organically, unfettered by a central command."[64] The ratio of distributed to traditional decision-making is reportedly about 70:30. Remarkably, the heads of all business units share responsibility for each other's success. Executives are compensated based on how well the collective of businesses, rather than individual businesses, performs. Employee compensation is based in part on their collaboration performance. Web 2.0 expert Ulrike Reinhard says that she can "feel the difference" when working with Cisco compared to more traditional companies.[65] For example, "when a decision needs to be made, the Cisco employee is able to act. But a comparable person in another German corporation has to go to someone else. Also, in proposing a video interview at Cisco, I get the response, 'Let's talk.' But, in other companies, the communication department has to go over it. It slows you down."

Embracing cutting-edge Web technologies

Integrated with these structural changes is a thoroughgoing embrace of cutting-edge Web technologies. Vice President and General Manager of the Software Group David Bernstein says (Sankar and Bouchard 2009: xvii): "Web 2.0 is our 'bet the company' strategy." The company sees its ability to "anticipate and prepare for market transitions (as) critical to Cisco's success and the success of its customers." In the company's view, "the Internet isn't a network of computers; it's a network of billions of people worldwide" (ibid.: 233). This strategy is enabling Cisco to "unleash the power of the 'human network effect' both inside and outside the company." For example, Cisco is using Web 2.0 technologies (e.g., *Cisco TelePresence*, *Cisco WebEx* and *Unified Communications*) to enable collaboration between employees, partners and customers. The company claims not only that using these technologies results in deeper relationships, but also that their use is "yielding increased productivity."

Greener, faster, cheaper meetings

In February 2009, Cisco held a virtual global company meeting using "Cisco TV," the company's internal video channel. That same year, CEO John Chambers also used Cisco's TelePresence to meet with international customers, radically reducing travel time (ibid.: 54): "In the future, consumers will leverage the visual networking capability of TelePresence, part of the media-enabled connected home, to interact with friends and family members across the country or around the globe—talking, sharing special events or even watching sporting events together." Even the cartoon character Dilbert now uses TelePresence (ibid.: 55).

The company employs social-networking and other Web 2.0 technologies (e.g., virtual reality) to increase collaboration among employees and transform business processes. Adoption of technologies such as discussion forums, wikis and WebEx Connect are leading to a company that is more collaborative and connected internally. Cisco is also leveraging new technologies to interact with its customers by using approaches such as "Digital Cribs" (ibid.: 262).

Cisco's collaborative channel

The Cisco platform aims to create a "collaborative channel," which is defined as "an ecosystem of collaborative relationships, rather than just a one-way pipeline for products or services." These relationships tie Cisco to its many partners, partners to end-customers, and partners and customers to each other. Thanks to this platform, the company now sees itself as just one of many nodes within a much more expansive universe of business relationships. Technologies such as WebEx, Cisco's Unified Application Engine (CUAE), the Cisco Compatible Extension (CCE) program for mobile and wireless devices, and AXP, a new Linux-based development platform for integrated services routers, are all examples of open platforms that allow partners to join a technology-development program of over 400 developers and co-innovate new applications and offerings on top of Cisco products (Tapscott 2008b: 11). In this ecosystem, partners can provide significantly more value than simply reselling Cisco products. For example, partners may deliver network platforms for data centers or unified communications in the form of an end-to-end managed solution that incorporates components from various manufacturers, providers and other channel partners. Partners can also integrate hardware and software from any number of vendors and wrap their own services around them. Cisco believes that the company and its partners are in a unique position to apply Web 2.0 know-how and technology to expedite their customers' ability to reinvent their business models ahead of their competitors.

Cisco reports that its most successful partners are transitioning from "box installers" and "system implementers" to "solution providers," thereby bringing an integrated set of hardware, software, consulting and service capabilities to their customers. A striking example of this "peer production" is the InService Alliance in North America. This alliance of 28 companies in 54 locations utilizes the combined strength of its members to offer a complete set of solutions to over 30,000 customers. Regional technology integrators are able to capitalize on opportunities that fall outside their individual regions and areas of expertise.

German Corporations

CoreMedia AG is a global supplier of content-management software headquartered in Hamburg, Germany. It was founded in 1996 as a spin-off from the *University of Hamburg*. Although CoreMedia was recognized in 2004 (along with *BMW*) as a "best innovator" among German small and medium-sized companies,[66] CEO Sören Stamer still faced an "existential challenge," asking (Buhse and Stamer 2008: 133): "(W)ith new ideas and processes bringing change spreading at an exponential rate ... (and) fundamentally changing the rules of play ... how can one stay in control when the market, the technologies in use, global competition and society itself (are) changing at an ever-increasing rate?" Stamer decided that the company needed to "let go" of its traditional approach to management and embrace the social-networking culture of Enterprise 2.0. Since then, Stamer has initiated a cultural transformation that integrates many tools of the Web. An online platform features many forms of social media (e.g., wikis, blogs, tags, rating systems and microblogging tools). The platform helps maintain porous corporate boundaries, which enables its employees, customers and partners to exchange information and communicate. The company found that "the continual dialogue concerning requirements and technical possibilities generates both ideas and expertise for innovations"

The company relies equally on a culture that values softer technologies (e.g., open-space principles) as a way of coupling individual passion with company needs and opportunities. These principles are founded on the values of transparency, openness and non-hierarchical, team-based organizing. The CEO strives to set the tone by welcoming feedback and criticism, while the company strives to emulate the intense interconnectedness of the brain and achieve a collective intelligence that integrates all available pieces of information into a unitary strategy for the organization. The firm's traditional hierarchy has been replaced by a flexible, highly networked structure in which individuals provide collective input to steer the firm. Top management retains responsibility and makes the decisions, but it uses the firm's collective intelligence as it does so.[67]

The SAP Development Network:[68] SAP was founded 25 years ago by a former group of IBM engineers and is located in the small rural town of Walldorf, Germany. Despite these humble surroundings, the company has grown to become the fourth-largest software company in the world. It creates the big company-wide software applications that today's firms use to run most of what they do. SAP provides a compelling example of the power of loosening traditional organizational boundaries in favor of creating a network of stakeholders operating in an ecosystem of mutual exchange and support. As Hagel, Brown and Davison observe, this must have been "somewhat scary ... for managers who are used to controling what takes place" (Hagel, Brown and Davison 2010: 5). But the willingness to take such risks paid off. As they describe it (ibid.): "SAP's example shows that companies can influence and shape the direction in which the community goes—so that it creates the most value for all the participants—without over-controlling things."

The software industry started to go through a wrenching change earlier in this decade as it moved from large, complex, tightly integrated application software to much more loosely coupled modules of software embedded in server-oriented architectures. SAP, whose success had been driven by the previous generation of software, embraced this next wave of software architecture by introducing its *NetWeaver* platform in early 2003—software that fit on top of and around its existing enterprise applications, helping them talk to each other as well as to non-SAP applications. In doing so, however, SAP ran into a classic chicken-and-egg dilemma: The product's full potential wouldn't become apparent until customers began using it and discovering what it could do, yet customers might not adopt NetWeaver—which SAP was essentially giving away as part of its applications—until they could grasp its potential. Even NetWeaver's early adopters—typically among the most tech-savvy of its customers—were struggling with its basics. Nevertheless, SAP had neither the reach nor the resources to train and teach its entire customer base—let alone educate tens of thousands of systems integration consultants.

Hagel, Brown and Davison explain what happened next (Hagel, Brown and Davison 2010: 74):

> That's when SAP executive board member Shai Agassi came up with a great idea: Why not let all of SAP's customers, systems integrators and independent service vendors teach each other about NetWeaver, peer-to-peer, as they learned to use it? The result was the SAP Development Network (SDN), a broad ecosystem of participants in generating discussion forums, wikis, videos and blogs. In one fell swoop, SP went beyond the limitation of its own resources to access a broad network of talented and passionate participants who proved to be crucial to the platform's success. The SDN community grew quickly and powerfully, and as it did, SAP established NetWeaver with its customers and third-party vendors. SAP's Developer Network and its related ecosystem initiative have created a rich network of 1.3 million participants contributing to more than 1 million separate topics of conversation.
>
> SDN was a success in no small part because it provided ample opportunity for nearly everybody involved to become more productive in what they do. Independent software developers could improve their coding chops. SAP's in-house code-writers could learn more quickly which of the features they wrote worked for their users—and which did not. SAP itself could get a lot more value from its customer service staff: As the SDN began taking care of more routine and entry-level customer questions, SAP could then focus on more difficult ones.

The SDN is an example of what Hagel et al. call a "pull" platform (as opposed to more traditional "push" strategies). Such platforms tend to be modular and designed for the convenience of participants. They are "loosely coupled, with interfaces that help users to understand what the module contains and how it can be accessed" (ibid.: 76) Such platforms "can accommodate a much larger number of diverse participants" and are designed to handle exceptions. They tend to have "increasing returns dynamics," that is, "the more participants and modules the platform can attract, the more valuable the platform becomes" (ibid.).

A Sobering Example

Eurostar learned the hard way about the importance of being prepared to use the tools of the Web to cope with PR crises. The company had invested in social media for marketing purposes, for example, by tweeting special offers and information about destinations. But since it had no crisis-management plan, it was not equipped to deal with passenger frustration when five trains heading for the United Kingdom broke down in the Channel tunnel in June 2010. Passengers then vented their frustration on *Twitter*, and Eurostar hesitated in responding. At this time, the company was also the victim of brand hijacking in the form of a Twitter stream going by the name of @Eurostar_UK, which turned out to be a fake. Within 48 hours, Eurostar rapidly overhauled its use of social media and used Twitter feeds to update customers on delays, cancellations and refunds. However, the case illustrates that, as Martin (2010) puts it, using social media "is not just about crisis management; it's a full-time operation. ... Customers don't know (or care about) the difference between marketing and customer service feeds. All they want is answers, and the company needs to provide them" (parentheses in original).

How the Web is Impacting the Social Sector

Challenges and Opportunities

The need to demonstrate impact

Leadership in the social sector must grapple with a paradox. As Kanter puts it (2010h):

> There has been an explosion in (the) size of the nonprofit sector over the last 20 years, huge increases in donations and (the) number of organizations, and yet (the) needle hasn't moved on any serious social issue. Growing individual institutions ... (have) failed to address complex social problems that outpace the capacity of any individual (organization) or institution to solve them.

The author's recommended solution for this paradox is evident in the title of her recent book, *The Networked Nonprofit: Connecting with Social Media to Drive Change*, coauthored by Alison Fine (Kanter and Fine 2010). A key message is that "nonprofits need to become more like networks and leverage the power of social media and connectedness."

Such awareness has been slow to dawn in the social sector. A 2008 study concluded that very few social-sector organizations were taking full advantage of what networks can do with new technologies. It found a "need for a baseline understanding" that was particularly strong regarding social media tools and a "heavy reliance on more traditional communication platforms, such as e-mail listservs, document sharing tools and the telephone" (Monitor Institute and David and Lucile Packard Foundation 2008: 3). However, due to the advocacy of Beth Kanter and a growing number of players in the social sector, Web 2.0 is gaining momentum and even leading the charge in some areas. It is becoming more and more evident that Web 2.0 has enormous potential for the transformation of organizations—and organizing—on behalf of social change.

Early adoption in philanthropy[69]

That social-sector organizations are beginning to draw on Web-based tools to make fundamental changes in the way they do business is most dramatically evident in the area of philanthropy. Although there has been debate over the usefulness of social networks for raising funds versus merely raising awareness (Hart and Greenwell 2009), a recent study makes a compelling case that the Web is transforming philanthropy in ways that reflect a larger transformation of the social sector (Bernholz, Skloot and Varela 2010: 4):

> Information networks—the Internet primarily, and increasingly SMS (text-messaging) and 3G (smart-phone) cell phone technologies—are overturning core practices of philanthropic foundations and individuals. Enormous databases and powerful new visualization tools can be accessed instantly by anyone, at any time. A decade of experimentation in online giving, social enterprise, and collaboration has brought us to a place from which innovation around enterprise forms, governance, and finance will only accelerate.

As recently as at the turn of the millennium, typical philanthropic activity consisted of some combination of volunteer work and donations to a nonprofit organization, whether it be a local one or a prominent national or international one. But now transactional philanthropy sites facilitate direct individual giving (as well as lending) independent of geographical location. Bernholz et al. found that (ibid.: 13):

> Today, individuals can lend money to small business owners in Tanzania, learn about the leanest, closest-to-the-ground nonprofits in Haiti, or buy art supplies

for a fourth-grade teacher in a rural school half a continent away. While it's true that, in the case of the Haitian earthquake, for example, most Americans donated to the *American Red Cross* rather than seeking out indigenous Haitian nonprofits, the trend is clear: With each passing year, more people learn about alternative candidates for their charitable dollars, in fuller and more revealing detail. In 2008, online giving surpassed $15 billion dollars (more than 5% of total giving), and in 2009, while foundation giving fell by a record 8.4 percent, online giving rose by 5 percent.

Discovering new ways to organize

There are at least four modes of civic engagement (Kearns 2005):
- Direct engagement (individuals acting alone);
- Grassroots engagement (individuals acting as part of a loose coalition);
- Network-centric engagement (individuals acting as part of a coordinated network);
- Organization-centric engagement (individuals working through social-sector and advocacy organizations with a governing board and centralized leadership).

Although only one of these takes place through organizations, that has still been the dominant mode in the past century. For many years, there has appeared to be a choice between getting things done by the state or by businesses, with foundations filling the gap in between. Almost exactly 100 years ago, Andrew Carnegie and John D. Rockefeller established centralized, vertically integrated foundations modeled on the big businesses that had given them their fortunes (steel and oil, respectively). This institutional structure has remained the predominant model for organized philanthropy for almost a century. The tacit assumption has been that people could not self-assemble. But, these days, electronic networks are enabling novel forms of collective action because they have made assembling so much easier. Now it is possible to have groups that "operate with a birthday party's informality and a multinational's scope" and that emerge through "ridiculously easy group forming" (Shirky 2008: 54, quoting Seb Paque). As a result, we are beginning to see the other three models of civic engagement.

Peer-supported, data-informed, passion-activated and technology-enabled networks are coming to represent an important new structural form in philanthropy. The institutions that support them will need to be as flexible, scalable and portable as the networks they serve. This shift to a "networked information economy" has enabled "the rise of effective, large-scale cooperative efforts—peer production of information, knowledge and culture" (Benkler 2006). One study concludes (Bernholz, Skloot and Varela 2010: 5): "On the cusp of the first modern foundation's centennial, we may be looking at the dawn of a new form of organizing, giving and governing that is better informed, more aware of complex systems, more collaborative, more personal, more nimble and, ultimately, perhaps more effective."

Network-centric engagement is a hybrid of the individual flexibility of the two less-structured forms of civic engagement (individual action and loosely formed grassroots networks) and the organizational efficiencies of the fourth. Its potential is radically increased by Web 2.0. Network-centric advocacy focuses on enabling a network of individuals and resources to connect on a temporary, as-needed basis to undertake advocacy campaigns. Leadership of campaigns can be decentralized. Furthermore, Web 2.0 makes possible the "just-in-time" delivery that has revolutionized manufacturing and retail in the business sector.

Network-centric organizing is especially visible in advocacy efforts, where organizations have been a constraint. Kearns argues that the "progress of the environmental movement has been stalled and in some cases reversed" by organization-based advocacy, which has been successfully countered by opposing organizations (Kearns 2005: 3). Advocacy organizations are also vulnerable to the "organizational dynamics of self-preservation, governance maturity, brand protection and specialization" (Kearns and Showalter n.d.). The challenge is to launch new forms of advocacy. In the United States, this would mean linking the 3,000 social sector organizations with organizations from other sectors, individuals and loosely organized teams through networks. This new form of advocacy is a way to counter the trend away from "joining" toward a more casually connected base of support.

Other research points to the benefits of network-centric vs. organization-centric ways of organizing. Research at the *Harvard Business School* suggests that organizational growth "does not necessarily translate into greater social value creation" (Lagace 2005). The author found that "the work of nonprofits is even more conducive to network forms of organization ... because the issues these organizations are trying to solve are large, complex problems that can't be addressed by any single entity" (ibid., citing Wei-Skillern).

For many tasks, "nonmarket entities and the self-organizing commons can compete with, and even outperform the market because market players tend to have higher overhead costs in the form of advertising, talent recruitment, capital equipment, attorney fees and so on. Funders can apply tremendous leverage by making relatively small investments in maintaining the infrastructure and information resources that enable nonmarket players to exist and flourish" (Bernholz, Skloot and Valera 2010: 36). Staffing individual foundations may cease to make sense, especially in the case of small foundations. Instead, "consortia of active donors may begin to thrive, especially for place-based or thematic endeavors, boosting the case for donor engagement in philanthropy" (ibid.: 37).

The need for new modes of governance

The new networked information economy also raises a fundamental definitional question: What is a legitimate social-sector entity? Network-enabled volunteer groups are emerging that are radically different from incorporated enterprises with bylaws, mission statements, formal boards of directors and geographical limits. They operate outside the existing regulations for grant funding that require nonprofit organizational status, they are managed by individuals who seek social solutions rather than monetary gain or market success, and they rely on new models of accountability. An example is Ushahidi, described in the examples later in this chapter.

Bernholz et al. conclude that the shift to such organizational models will require new modes of governance (ibid.: 2010: 38):

> Most of the successful examples of distributed governance, such as the ongoing development of the open source software platform Linux, are made possible by norms and licenses that are unique to the software arena. For other kinds of ventures—such as in higher education, medical research or service provision— where open source content sharing is not a norm, rules of the road for governing networks and networked organizations may need to be invented. ... The reconfiguration of business forms and the development of hybrid governance models will undoubtedly stress the laws, regulations and cultures that have developed in isolated silos. We will not only see the blending of market and nonmarket organizations; we will see the corresponding development of new approaches to funding, finance, and reporting requirements.

This study found that the social sector has its own version of the business world's IP issues (ibid.: 32):

> Who ultimately *owns* social-sector data is an unresolved issue for donors and enterprises. ... In the public sector, research funded by the National Institutes of Health must be published in the openly accessible PubMed database within 12 months of work completion. In the case of philanthropy, because donors receive tax benefits—essentially, unrestricted grants from the government— foundations are quasi-public institutions. Data held by foundations would therefore seem to belong to the public. Most foundations don't behave as if they, or the data they produce, are owned by the public. While a few funders have become more open by publishing grant applications on their websites, information about selection and performance rarely sees the light of day.

The rise of free agent activity

The option of civic engagement initiated by individuals has become far more viable over the past decade. Kanter and Fine have labeled those producing this powerful new force "free agents," defined as "individuals working outside of organizations to organize, mobilize, raise funds and communicate with constituents" (Kanter and Fine 2010: 15). Some work on behalf of nonprofit organizations, while others work independently. Often they make it possible for other individuals to operate more independently. For example, 10 years ago, "individual citizens were unable to contribute directly in response to a natural disaster like the 2001 Gujarat, India, earthquake. The best they could do was send money to a large international nonprofit like the *American Red Cross*. Today, a worldwide community of 'crisis mappers,' using satellite imagery and on-the-ground information reported via cell phone, helps coordinate responses to complex humanitarian emergencies" (Bernholz, Skloot and Varela 2010: 4). Likewise, a decade ago, "microfinance was entirely top-down—from large institutional lenders to small borrowers. Today, anyone can lend $25 to entrepreneurs located anywhere on the globe" (ibid.).

The need to share ownership of information

The Web has also brought to the fore the issue of ownership of information, the social sector's equivalent of intellectual property concerns. The widespread availability of broadband Internet access and the near ubiquity of SMS and 3G cell-phone networks has given everyone the tools to both produce and consume. These have expanded individuals' sense of empowerment and led to profound changes in expectations and norms. What information matters to funders and nonprofits? Who has it? Who owns it? How do we share it? How do we collaborate around common issues? It requires the top team to actively create both a culture and a process for proactively allowing their knowledge to be reused in the correct way. Most nonprofits would do well to have a policy around the use of "Creative Commons" licensing. A 2009 study performed by the Berkman Center for the Internet and Society, at the Harvard Law School, found that while "open licenses promise significant value for foundations and for the public good, and often for grantees as well" (Malone 2009: 46), they are rarely used in the philanthropic sector because "many grantees and foundations are relatively uninformed and inexperienced with open licenses" (ibid.:41).

Following are examples of pioneering responses to the overwhelming challenges and changes faced by the social sector.

Patterns and Examples

Internal Communication and Organization

Improving internal communication and knowledge management

Paul Levy, CEO of *Beth Israel Deaconess Medical Center*, created a public blog to "share thoughts with people about my experience here and their experiences in the hospital world."[70]

Through an internal blog, staff members at *ZeroDivide* document lessons learned as the foundation implements a new grant-making program to support social enterprises (Luckey, O'Kane and Nee 2008: 3).

The *Open Society Institute's* "KARL" uses wikis, blogs, tagging and other Web 2.0 tools to enable communication and collaboration among employees separated by culture, geography and program boundaries. KARL also facilitates communication between the organization and its grantees and partners.[71]

Alternative forms of organization and governance

Ushahidi "was started by an unincorporated group of colleagues spread over two continents and several countries. Even though the informal, networked structure proved capable of building an effective platform for the advancement of social good, that same structure proved to be a stumbling block for raising foundation funds. It didn't conform to the organizational model funders understood and were comfortable with" (Bernholz, Skloot and Varela 2010: 26).

Linking people across geographically dispersed organizations

At *Catholic Healthcare West* (CHW), more than 60,000 caregivers and staff members deliver care to diverse communities across Arizona, California and Nevada. As one of the largest hospital systems in the United States, CHW has a unique challenge—uniting 41 separate facilities and their employees under one common brand and cause. Their solution is to utilize "employee-generated content" (ECG), which tells the stories of how each person at CHW "lives their life on purpose" (Center for Creative Leadership 2008).

Closer Conntection Between Organizations and Expertise

Crowdsourcing

In the social sector as well as the business sector, *InnoCentive* is a potential broker for crowdsourcing solutions to R&D challenges. As noted above, it creates an innovation marketplace connecting companies and academic institutions seeking breakthroughs with a global network of more than 125,000 scientists, inventors and entrepreneurs. The Rockefeller Foundation is allowing nonprofits to use the InnoCentive process to post problems related to addressing the needs of poor and vulnerable populations and offering rewards to innovators who solve them. As also noted earlier, corporate users of InnoCentive include *Alcatel-Lucent, Siemens* and *Unilever*, and organizations similar to InnoCentive are *NineSigma, Elance* and *YourEncore* (Tapscott and Goodwin 2008: 9).

The *David and Lucile Packard Foundation* is using a public wiki to gather insights from stakeholders to inform its grant-making strategy related to nitrogen pollution.[72]

The *Omidyar Network*, a philanthropic investment firm launched by *EBay* founder Pierre Omidyar, asked the public to participate in awarding its grants. It created an online framework for the interested community to deliberate on and winnow down the proposals (Noveck 2008: 2).

The *Global Greengrants Fund* makes small grants to grassroots environmental groups working around the world, and uses a network of regional and global advisory boards to make region-based decisions.

Ashoka's Changemakers builds communities that both compete and collaborate to find solutions to social problems. Expert judges select a set of finalists, but the final winner is chosen through a vote by members of the online Changemakers community (ibid.).

Paul Buchheit, an early *Google* employee, wrote a blog post looking for advice on his donor-advised fund and then built a series of online tools—including a Google-moderated voting site and a *FriendFeed* group—that enable anyone to post suggestions. The *British government* proposed a similar project to guide some of its funding for international aid (Bernholz, Skloot and Varela 2010: 22).

The *John S. and James L. Knight Foundation* has used crowdsourcing tactics in its News Challenge grants program, and in 2007, the *David and Lucile Packard Foundation* used a wiki to solicit possible approaches to dealing with the problem of nitrogen pollution (ibid.: 22–23).

Opening up goal-setting and strategy formulation

- The *Lumina Foundation for Education* "has posted its strategic planning process, the plan itself and the progress measures being used on an interactive website to which the public can contribute comments. The foundation also has a *YouTube* channel on which the public can watch and comment on video interviews with key decision-makers" (Bernholz, Skloot and Varela 2010: 22).
- The *Peery Foundation*, in Palo Alto, California, "recently pushed its strategic planning conversations into public view using *Twitter*—welcoming thoughts, sharing its planning tools and actively discussing its ideas with anyone who followed the foundation's board or staff members. The Twitter discussions prompted prominent bloggers to weigh in on the process" (Bernholz, Skloot and Varela 2010: 22).
- The *Smithsonian Institution* has used a wiki to crowdsource its strategic plan (Kanter 2009c).

Fostering Openness

Facilitating public access to information

- Led by volunteers and managed remotely with free software, *Nonprofitmapping. org* rates states "on the quality of data on nonprofits they make available, with the aim of improving state reporting standards. It's an example of how a virtual team, without an organizational home, permanent institutional affiliation or shared locale, can work together to solve a big problem" (Bernholz, Skloot and Varela 2010: 38).
- In its own words, the *Sunlight Foundation* "uses the power of the Internet to catalyze greater government openness and transparency, and (it) provides new tools and resources for media and citizens alike." It creates new tools and websites to enable individuals and communities to better access information and put it to use. For example, in the arena of campaign finance, the foundation "enables users to tease out who gives how much money to whom, when they give it, and (by implication) why" (Bernholz, Skloot and Varela 2010: 39).
- The work of the *Milken Foundation, FasterCures*, and a few other philanthropic bodies points the way toward a future of greater access to important information for the public good (ibid.: 32). (See the FasterCures case profile below.)
- Voluntary efforts such as the *Public Library of Science* and *Science Commons* have laid the groundwork for sharing information in pursuit of common goals (ibid.: 32).

Building Ecosystems of Support

Connecting grantees with peers and experts

- The *MacArthur Foundation* has hosted online discussions between clusters of grantees and issue experts (Luckey, O'Kane and Nee 2008: 3).
- The *Nonprofit Technology Enterprise Network* (NTEN) provided educational forums for its members via webinars led by issue experts in the field of nonprofit technology. By using a website plug-in (*Gabbly*), NTEN provides attendees with a chance to hold "back channel" chats during the seminar. A podcast of the session is posted on the organization's website (ibid.).
- The social-sector technology hub *TechSoup* created *NetSquared* (Net2) to help social-sector organizations learn about and utilize social Web tools.

Sharing knowledge and interacting with the wider community

- In 2008, the *Meyer Memorial Trust* created *Connectipedia.org* to serve as a collective intelligence wiki space for the social sector.
- The *Natural Capital Institute* created *WiserEarth.org* in 2005 to "identify and connect the hundreds of thousands of organizations and individuals throughout the world working in the fields of environmental sustainability and social justice."
- The *Skoll Forum on Social Entrepreneurship*, a popular annual event that brings together leaders in the field of social entrepreneurship, has become interactive, with real-time chat and streaming video (ibid.: 2).

Attracting Resources

Fundraising[73]

- *Tweetsgiving*, a campaign launched by *Epic Change*, raised over $10,000 in 48 hours to build a school in Tanzania.
- *Twestival* raised $250,000 in three months for "charity: water" by sparking local "Tweet-ups" for giving.
- Social networks enabled people concerned about the 2010 earthquake in Haiti to generate an unprecedented amount of both money and expertise in a remarkably short period of time. As Bernholz et al. write (Bernholz, Skloot and Varela 2010: 24): "Many used social networks to spread word of the disaster, round up funds and volunteers, and stay informed about developments in Port-au-Prince. To date, more than $1 billion has been collected for relief and reconstruction, with the average donation via the Internet at a mere $10."

71

- Websites such as *GlobalGiving*, *DonorsChoose* and *VolunteerMatch* "facilitate donations of money or labor. Other sites, such as *Kiva*, *MyC4* and the *Social Impact Exchange*, facilitate loans and other forms of investment. Online information hubs such as *GuideStar*, *GiveWell* and *Charity Navigator* describe and assess the quality of nonprofits. The last category represents a genuinely new development in the philanthropic landscape. These sites can potentially connect a vast number of potential donors (institutional and individual) to a vast number of potential recipients" (ibid.: 8–9).

- Similarly, volunteer-driven "flash" causes can create tremendous impact by drawing attention to an issue for a very brief period of time. As Bernholz et al. write (ibid.: 38): "Some can even move a fair amount of money. In February 2009, 'charity: water' raised hundreds of thousands of dollars through parties in more than 100 cities, all organized by volunteers via *Twitter*. These dispersed, crowd-organized events are common tools of community organizing and political fundraising and are increasingly present in campaigns for charitable support."

- *X Prize Foundation*, an example of an incentivizing prize competition, designs a competition so that participants spend more money cumulatively than is offered as a prize. "The prize challenges extended by the *Rockefeller Foundation* and administered by *InnoCentive* draw upon the talents and expertise of individuals who might not otherwise devote their time and energy to solving problems in the social sphere. More conventional prizes, awarded on the basis of merit, include *Ashoka's Changemakers*, the *MacArthur Foundation's Digital Media and Learning Competition*, and the *Case Foundation's 'Change Begins with Me' challenge*. All engage new types of partners in both discussing issues and developing solutions" (ibid.: 23).

Creating individualized ways to donate

- Sites such as *Social Actions* or *All for Good* pull together and make available multiple donation or volunteer opportunities in a given locale or on a certain issue (ibid.: 14).

- Kiva.org, a microlending site, allows people to easily lend money to the working poor. So far, according to Kiva's reports, some 520,000 people have loaned more than $80 million to people in 184 countries. Using *PayPal* or a credit card, a visitor to the *Kiva* website can loan a struggling entrepreneur in a developing country $25 or more. The site says the money is usually paid back within a year. Other microlending sites include *DonorsChoose* and *GlobalGiving*.

- *Germany's betterplace.org*, operated by the *betterplace Foundation*, enables people to seek support for their own initiatives or to find and support projects that are meaningful to them. The foundation vows to ensure that 100 percent of donations are channeled to the intended targets. Overhead is covered through the profits made by the related company betterplace Solutions GmbH and its partners.

Blending donations with investments

- Web tools are enabling a blending of online giving markets that manage charitable donations and investor-level exchanges that manage social investments. "In some cases, such as the Denmark-based site *MyC4*, the user determines on a case-by-case basis whether she is making a gift, a loan or a profit-seeking investment. Other sites, such as *Kickstarter*, which supports artistic and cultural projects, acknowledge that the funds they drive to projects can be classified as investments, gifts, loans or any combination of the above—leaving the decision to the funder and recipient and broadening the options of both" (ibid.: 14).

Enabling Rapid Response

Responding rapidly to crises

- The *Ushahidi* platform enabled the *International Network of Crisis Mappers* (CM*Net) to respond to the 2010 Haiti earthquake by "rapidly gathering and disseminating information on the location of safe water resources, disease outbreaks, fuel sources, and hospitals and medical aid stations. CM*Net-produced data were used by the U.S. military, the Haitian government and dozens of nonprofits in planning and coordinating their response" (ibid.: 25).
- Bernholz et al. write that, in the aftermath of the major devastation in Haiti caused by the 2010 earthquake, "we have seen how quickly and on how large a scale individuals and organizations can collaborate on behalf of others. In a matter of days, three platforms—text donations, *Twitter* and *Facebook*—moved from the philanthropic margins to the center of both fundraising and volunteer activity. A series of loosely managed, globally dispersed weekend "CrisisCamps" took place on several continents over many weeks. Volunteers at these events produced dozens of software tools to help relief workers on the ground and in government agencies" (ibid.: 24–25).

Mobilizing people rapidly

- "Smart mobs"—large groups of people linked by cell phones, text messages, e-mails or other technologies—can assemble suddenly in a public place to perform some collective action in support of a cause. This was demonstrated for the first time in 2001 in the Philippines to protest government corruption and help oust then-President Joseph Estrada (Rheingold 2003, as reported in Shirky 2008: 174–175). Such "flash activism" has since become a common strategy (Schwartz 2009).

An interesting variation is a so-called carrotmob, which is a network of consumers who buy products to reward businesses making socially responsible decisions.

Sounding crisis alerts and providing support

- Despite government restrictions on the media, Web-based photo-sharing provided up-to-the-minute documentation of the 2006 military coup in Thailand. *Wikipedia* served as a clearinghouse for information (Shirky 2008: 36–37).
- The *Katrina PeopleFinder Project* evolved to engage volunteer programmers in developing a single site that allowed people to search dozens of separate databases and message forums to find lost relatives after Hurricane Katrina (Tapscott and Williams 2006: 186–188).
- *U.S. State Department* employee Jared Cohen described the "Neda video" (of the young woman shot during the anti-government demonstration in Tehran in 2009, who died while being videotaped by a cell phone) as "the most significant viral video of our lifetimes" and told the site's senior management that *YouTube* is in some ways "better than any intelligence we could get, because it's generated by users in Iran" (Lichtenstein 2010: 27).

Building networked organizations rapidly in response to emerging problems

- *Voice of the Faithful* formed quickly in response to a series of articles in the *Boston Globe* about sexual abuse scandals in the Catholic Church. In less than a year, the organization grew from 25 local members to a 25,000-member global network (Kasper and Scearce 2008: 3).

Expanding Capacity to Serve

Extending services through affiliated organizations

- In 1990, *Women's World Banking* provided 50,000 women with microfinance services. Ten years later, it served 10 million, by fostering a network of affiliates and associates that were themselves independent organizations. The founder, Nancy Barry, suggests that "instead of thinking about management challenges at the organizational level, leaders should think about how best to mobilize resources both within and outside organizational boundaries to achieve their social aims" (ibid.: 2).

Working with free agents

- Kanter and Fine report that *Tyson Foods*, in partnership with other organizations and leading bloggers (e.g., Chris Brogan), sponsored the "Pledge to End Hunger." For every thousand people who signed the online pledge form, Tyson donated 34 pounds of food to a food bank. Free agents were encouraged to blog about the campaign and spread the word, which resulted in nearly 5,000 pledges (Kanter and Fine 2010: 19).
- *LIVESTRONG* embraced and promoted Drew Olanoff, who acted as a free agent to create a tongue-in-cheek website (www.blamedrewscancer.com) and a Twitter page to share his anger and humor while coping with a cancer diagnosis. Drew guest-blogged at LIVESTRONG, where readers were encouraged to donate a dollar per complaint (ibid.: 20).

Providing support for customers

- *Massachusetts General Hospital* uses *CarePages*, an online blogging system, to help patients with critical health issues manage the challenge of communicating the status of their health with family and friends (Li and Bernoff 2008: 153–157).

Using mashups for business intelligence (BI)

- The *University of Louisville's Pediatrics Foundation* implemented a mashup for flexible reporting in support of its operation of medical clinics and pediatric in-hospital services. The foundation adopted eThority's mashup BI system to enable its analysts to build customer reports from accounting data in *PeopleSoft* and *Sage PFW* (Mulholland, Thomas and Kurchina 2007).

Capacity-building

- *WeAreMedia* offers an online curriculum for social-media nonprofits created by over 200 social media and nonprofit practitioners.
- With the support of SSE (*School for Social Entrepreneurs*) and *Unltd* Level 1 & 2 awards, Nathalie McCermott set up OnRoadMedia.org in 2005 to help marginalized groups and volunteers in the United Kingdom use social media. On Road Media currently delivers training courses that take place on-site within an organization or community center. For example, it has trained homeless people through *CRISIS*, people with mental health issues through the *NHS* and young people at risk through *Catalyst Communities Housing Association*. The organization has also trained staff from *Oxfam*, *Unltd* and *The Prince's Trust*.

- Susan Mernit's *Social Media Tool Box*[74] is another example of a resource created to support social-sector organizations in using social media. It identifies useful steps in coming up with a social media strategy.

Access to knowledge

- Social media now enable new kinds of collaboration both during and after conferences. For example, "a 2009 conference on social capital markets devoted several weeks before the event to building up awareness on blogs and *Twitter*, had volunteers updating *Facebook* and *Flickr* pages before, during and after the event, showcased two different video channels, one live and one recorded, and equipped several participants with small video cameras to capture sessions as they happened" (Bernholz, Skloot and Varela 2010: 25–26). Both those who were at the conference and those not able to attend were able to access this information online. Training in the use of these new tools in workshops is already being offered (Kanter 2010b).

Providing citizens access to useful information

- *Safe2Pee.org* helps people find public toilets.
- *Freecycle Network* is a global network of local groups composed of volunteers who help link people in possession of unneeded but still usable items with others who might need them in the hopes of reducing waste.
- *Couchsurfing.com* helps people make connections with people living in places they travel to and find a place to stay.
- *FrontlineSMS*, an open source software program that "enables mobile-phone users to send text messages to large groups, has been used by local individuals and enterprises to post updates on commodity market conditions in rural Peru, report the location of landmine victims in Cambodia, and record human rights violations in Ghana" (Bernholz, Skloot and Varela 2010: 24).

Enhancing Effectiveness

Enhancing measurement of performance

- Donors and investors are also actively engaged in developing whole new systems for measuring progress. The *Acumen Fund*, "an independent social investment fund focused on alleviating poverty in Asia and Africa, has begun developing internal measures of progress that can be used across its portfolios. Each portfolio addresses a distinct domain, such as job creation, health outcomes or access to clean water" (ibid.: 27).[75]

- *Success Measures* has created a variety of Web-based evaluation framework designs. As Bernholz et al. write: "Groups can aggregate data, download them to Excel to create spreadsheets and graphs, and contribute to the further refinement of Success Measures frameworks, tools and indicators by sharing what they learned. To date, more than 300 community development practitioners, intermediaries, funders, researchers and evaluators have participated in the development of Success Measures" (ibid.: 27–28).
- The *Noaber* and *d.o.b. foundations* have created "Social e-valuator," a tool that enables organizations to determine the social return of a project or program. This interactive online tool enables an organization, social enterprise or business to shape a project, continuously changing factors to result in the highest possible social return. This is an example of a "public dashboard" (Kanter 2009e).

Improved data gathering

- *GlobalGiving* illustrates how networked technology can reduce the expense of obtaining on-the-ground data. "Working in a small African village, the group's leaders handed out bumper stickers that asked people to text their thoughts about the program to a certain number. Anyone with an opinion could respond anonymously about the impact, management and role of the organization in the community" (ibid.: 28).
- *Keystone Accountability*, a U.K.-based research and consulting firm, has pioneered more sophisticated techniques for data collection. The firm "offers a free tool on its website to enable nonprofit organizations to acquire anonymous constituent feedback" (ibid.: 28).
- An evaluation of *YouthTruth* (a partnership between the *Bill & Melinda Gates Foundation* and the *Center for Effective Philanthropy*) provides another example of the kinds of data that can now be collected through social media. "YouthTruth distributed a survey online (via *MySpace* and *Facebook*, via email and with the help of *MTV*) to high school students attending schools receiving funding from the Gates Foundation. The data collected are used to inform the schools, the funders and the evaluators" (ibid.: 28).[76]

Enabling transparency of communication with grantees and easing grant-making transactions

- At the *Skoll Foundation*, prospective award applicants can easily determine their eligibility through the online "eligibility quiz" (Luckey, O'Kane and Nee 2008: 3).

- *DonorsChoose.org* is an online marketplace for connecting donors with opportunities to support public schools (Monitor Institute and David and Lucile Packard Foundation 2007: 9).
- The *Charles Stewart Mott Foundation* has a searchable online database. An RSS feeds of grants awarded makes grant-making data available to all.

Reducing Waste and Fraud

Cutting costs

- The cost of processing an e-payment is about two cents, compared to 43 cents for issuing a check (Eggers 2007: 29).
- The U.S. state of Florida reduced the number of employees delivering HR services by more than 1,000 positions (ibid.: 30).
- If the government can replicate the private sector's 20 percent average savings from putting processes online, it has been estimated that all levels of the U.S. government could together save $100 billion annually (ibid.: 32).

Reducing fraud

- Eggers reports a variety of ways in which "new technologies have proven highly effective in rooting out tens of billions of dollars in taxpayer money that's wasted on fraud, abuse and erroneous payments" (ibid.: 31). For example, the Canadian province of Ontario saved a projected $1 billion over five years by discovering that 17 percent of all welfare recipients were ineligible and another 8 percent were overpaid.

Fostering Individual Engagement

- *MediaVolunteer.org* organized the time of nearly 20,000 volunteers to develop a media contact database for progressive organizations.
- The online news site *Muckraker* asked its readers to make sense of the 3,000 e-mails released by the *U.S. Department of Justice* related to the firing of federal prosecutors in 2007. Within hours, readers were identifying questionable passages and thereby leading to new story leads (Kasper and Scearce 2008).

Expanding participation

- It's now possible to barter for or donate goods simply by posting on *FreeCycle* or *Craigslist* (Bernholz, Skloot and Varela 2010: 14).
- New cause-oriented sites, such as *Causecast* (which helps people find causes to support) and *Amazee* (which showcases various social-advocacy projects) are increasingly common.

Providing incentives and support structures for volunteer contributions

- *Timebanks.org* has created a system that "connects unmet needs with untapped resources." It does so by using the soft currency of contributed time to reward participants who volunteer their skills by enabling them to trade their accumulated credit for access to skills contributed by others.
- *The Extraordinaries* is a pioneer in the new field of "microvolunteering," linking volunteers with a mobile phone and a few minutes to spare to organizations in need of assistance. (See the more detailed case description in the following section.)

Cases in the Social Sector

FasterCures: Believing that medical research was often conducted inefficiently, even counterproductively, and that funders were part of the problem, Michael Milken founded FasterCures and the FasterCures *Philanthropy Advisory Service* (PAS). These organizations aim to change the way research institutions and funders develop and share knowledge. FasterCures performs independent research on a variety of diseases and disease research institutions. The research is made available on the Web through the PAS, whose members have access to reports on diseases and searchable disease databases. In their excellent profile of FasterCures, Bernholz et al. write (Bernholz, Skloot and Varela 2010: 16):

> The PAS marketplace increases funder efficiency by steering donations toward research on those diseases whose cures appear to be closest to breakthrough and toward those institutions that score highest on assessment reports. It improves entire disease research fields by motivating institutions that receive poor assessments to improve their practices. It also eliminates the need for each PAS member to separately perform due diligence on multiple potential grantees, thereby solving one of the "reinventing the wheel" problems that continually plague organized philanthropy.

79

FasterCures is also having an impact on organizations that conduct disease research, which are now able to "benchmark themselves against a set of independently generated and tracked standards, report their results against consistent parameters and organize their work in new ways" (ibid.). They also come together "to share ideas on knowledge development, organizational practices, community engagement and research—so that if experts working in one disease arena have a breakthrough, the process of others' learning from the breakthrough and applying it can be accelerated" (ibid.). A network of "cure entrepreneurs" has emerged, which can "move innovative solutions across formerly siloed institutions and disease communities." Bernholz et al. conclude (ibid.):

> FasterCures is one example of a foundation-led effort to transform how both donors and doers work. It's built on the premise that donors will value in-depth analysis of a field and of the organizations engaged in it; and that competition, made possible by a networked information marketplace, can improve efficiency in whole fields.

The American Red Cross: Wendy Harman's experience at the American Red Cross illustrates both typical organizational barriers to using social media and the success—and value added—that can result from creativity and persistence. Facing strong criticism for its lackluster emergency response to the impact of Hurricane Katrina, the Red Cross hired Harman as its first social media manager. On arrival, she had to overcome the security-conscious IT department's blocking of access to sites such as *MySpace* and *Facebook*. She was not able to get approval to start a blog or have a *Flickr* account to show volunteers doing their work. She simply went ahead without permission, using a personal credit card to purchase a domain name and create accounts. According to Charlene Li, Harman "addressed with persistence and patience each concern and fear her executives had about engaging in social media, from malware downloads to confidentiality of clients shown in pictures uploaded to *Flickr*. She made sure the proper processes and procedures were in place before entering each new media channel" (Li 2010: ix–xi). Over a period of two years, she added a blog, Flickr, Facebook pages and *Twitter* accounts. To address the problem of potential inconsistency in Web usage among the more than 700 local chapters of the Red Cross, she wrote a handbook laying out guidelines, procedures and best practices for how the chapters should use social media. Her efforts not only prevented problems, but also began to generate unanticipated results. The large base of people who are employees, blood donors and responders became part of the organization's outreach. The retailing giant *Target* ran a Facebook-based fundraising contest for selected organizations, including the Red Cross, which generated $749,000. Once Harman had demonstrated success, her supervisor granted her request for reimbursement (ibid.: 149; 175–176). The enhanced capacity of the organization was demonstrated during its Haiti earthquake response in January 2010, when it raised over $10 million in three days, largely due to easy donation channels on Facebook and Twitter. Li observes (ibid.: xi): "What's fascinating about this story is that the American Red Cross started engaging in social media because it sought to control it, but realized over time that it was better to be open and engage with those who were already engaging *them*."

The Bertelsmann Stiftung:[77] In 2008, the Bertelsmann Stiftung's "Shaping the Global Future" team began work on developing a new Internet platform designed to offer a forum that would engage with the wide array of networking opportunities opened up by Web 2.0. The platform's aim is to promote a dialogue between experts and non-experts focused on the correlations between global megatrends such as climate change, demographic change, migration and pandemics. The idea behind the platform, www.futurechallenges.org, was to enable an international exchange of views on the extremely complex issues humankind will face in the future, and at the same time, to nurture mutual understanding between people in different regions of the world in their dealings with the impacts of these mega-trends at their own local level. In its use of the Internet, the Bertelsmann Stiftung was also attempting to reach target audiences that have not been accessible via classical communication channels. The platform recorded a total number of around 800,000 hits in the first five months following the launch of its beta version and has recruited 50 regular bloggers from 33 countries. Since its launch, the website has also been (re)presented at some international conferences, enabling its bloggers to stream live reports from them.

As the FutureChallenges project developed, it increasingly took on the character of a venture into new and unexplored territory. This was also the case in terms of leadership because, in 2008, almost no one within the executive team itself had much experience in dealing with Web 2.0. Initial considerations regarding an overarching Stiftung strategy for working with it started at that time. The increasing reliance on social media tools (e.g., Facebook, Twitter and Xing) and various (internal) wiki and blog applications—which were essential for building an international pool of bloggers who would regularly contribute to the new website—had its impact on the management of this project. Use of such tools demanded both a new kind of transparency and a willingness to accept the greater uncertainty that inevitably accompanies innovation and experiment. This became particularly clear while setting up and managing an international blogger network, which is organized into seven groups corresponding to different regions of the world (Africa, Asia, Middle East, America, Europe, etc.). For each of these regions, a so-called regional editor is responsible for leading a group of 10–15 bloggers. By deciding to have these editors, the team was basically deciding to decentralize the content-generating process for the whole platform. This decision made it possible for the core team to concentrate much more on general questions about how to develop the platform further. However, for the sake of consistency, it was nevertheless necessary to also find ways to communicate directly with the bloggers on some issues to explain to them various general directions of the project and the foundation's point of view. A hard lesson for the team to learn was that authenticity on the Web could only be achieved if the team itself engaged in the different social media forums.

Given this, it took time and effort to reach a common understanding of what the project was about—that is, its strategies and objectives. The course of project development saw different modifications of the project's aims and objectives, which all necessarily entailed technical and agenda-related changes to the set of platform requirements. This in turn led to higher and more complex communication requirements imposed on the service provider charged with programming the website. On top of this, some additional Web 2.0 experts were brought into the development process after several key decisions about the technical implementation of the website had already been made. Looking back, it appears obvious that an early and ongoing dialogue between Web designers and active social media experts is an essential element in a successful development process in this field. In this case, the Stiftung also learned the importance of involving its own IT, communication and legal departments throughout the whole process so as to harmonize internal and Web 2.0 requirements. In practical terms, for example, this meant finding a legal solution to the use of a Creative Commons license for the content of the whole website or to such knotty issues as how to classify blogger postings for tax purposes.

A report on the project concluded:[78]

> FutureChallenges.org has been a great learning experience for us ... [We have learned] innovative management tools (e.g., Internet-based planning software), modular software development in parallel with the Internet launch of the platform, new communication requirements and the challenges for leadership in such flexible or even virtual environments. It is too early to judge whether the project will be successful, but we now have an idea of what it takes: all of our senses to feel the demands of our "customers," all our strength to establish and maintain a mutual understanding of the direction we are going in, and all our leadership capabilities to provide the guidance and stable framework needed to create the necessary space for innovation.

The Extraordinaries: Currently in development, the San Francisco-based "The Extraordinaries" is "part of a new movement that combines tiny technology and huge social goals." It is an online platform that seeks to make it easy for altruistic consumers to support an organization or cause they care about. Its potential has generated a number of awards, including: a $60,000 two-year fellowship from Echoing Green, a nonprofit group that awards grants to social entrepreneurial organizations; a United Nations World Summit Youth Award; and a $249,000, one-year John S. and James L. Knight Foundation community grant (Weeks 2009). It enlists both individuals and groups of company employees to contribute their expertise to a nonprofit in even the smallest chunks of time. Nonprofits begin by posting requests to the site, which are then routed to would-be volunteers based on their skills and interests. The Extraordinaries is in the process of becoming a so-called B Corp. Its business model will eventually include charging organizations a fee for each task completed.

Smartphone applications such as The Extraordinaries and *Catalista* allow people to donate mental labor wherever they are and whenever they like (den Toon 2010). Similar to the way in which Catalista connects would-be volunteers with opportunities by mobile phone, The Extraordinaries seeks to enable "micro-volunteering." Examples might include translating a page of a document into Spanish, helping to choose a new logo or offering advice on a college application. The Extraordinaries even has pre-built "kits" that turn a series of best practices into tasks for volunteers. Willing volunteers then complete the tasks during a spare moment via *iPhone* (through a dedicated app) or a Web browser, or they can share them with their colleagues. Either way, corporate team volunteers can track each others' efforts via a "Team Activity feed."

This novel approach is not without barriers. Organizations aren't accustomed to accomplishing tasks through crowdsourcing. Likewise, some volunteering opportunities (e.g., donating blood or making a sandwich for the homeless) don't lend themselves to smart phone apps. But, as Extraordinaries co-founder Jacob Colker told National Public Radio (NPR) in 2009, micro-volunteerism "is perfectly suited for the Millennial Generation. They are used to text messaging, *MySpace*, *Facebook*, get-in, get-out, instant gratification. For them, going out and cleaning up a park—that's not necessarily attractive to them. As we introduce them to the warm fuzzy feeling of doing good, that will increase awareness" (Weeks 2009).

How the Web is Impacting the Government Sector

Challenges and Opportunities

Overcoming bureaucracy

Of all the traditional sectors, the public sector poses the greatest challenges of all because it must cope with the full gamut of societal threats and ills. Yet in even the most well-developed democracies, it must meet these challenges within severely limiting constraints. The most fundamental is its form of organization: bureaucracy. Although bureaucratic organization was a revolutionary innovation when originally introduced in the late 18th century,[79] many current observers would concur with the assessment by William Eggers that "a bureaucracy built for the Industrial Age can't adapt to the Age of Information." He notes that, in contrast to the private sector, public bureaucracies, "with their vertical information flows, rigid practices and strict division of labor, are still organized according to the top-down models created for the industrial economy" (Eggers 2007: 2). Additional constraints on the ability of governments to respond to their increasing challenges include IT obsolescence, the complexity of the political process, corruption and influence-peddling—all of which are exacerbated by declining levels of resources.[80]

Web 2.0 is widely seen as having a huge role to play in overcoming many of these limitations and as a resource to meet the challenges in new and creative ways. For this reason, the term "Government 2.0" (or "Gov 2.0," for short) has recently been coined. Moreover, the Web has rejuvenated the "open government" movement, which had aspired for decades—with little success—to make governments throughout the world more transparent. In the United States, this movement was set back even further by the prevalence of "closed government" practices during the administration of George W. Bush (2001–2009).

Despite this great need and strong potential, the public sector in countries across the globe generally lags far behind in responding to the potential of Web 2.0. Not only does it (understandably) trail behind the business sector (ibid.: 29); in many ways, it also trails behind the social sector as well. As recently as the spring of 2009, one commentator wrote (Pople 2009): "Most (government) experiments with Enterprise 2.0 have been limited to the creation of additional communication channels to broadcast organizational messages to a defined customer set, albeit with good value in speed and targeting of delivery." Although this observation was specific to the United States, there is no reason to doubt that it would apply equally well to most if not all governments across the globe.

Rising prevalence and sophistication of Web 2.0 usage

The first wave of digitally enabled "e-government" strategies delivered some important benefits. It made government information and services more accessible to citizens while creating administrative and operational efficiencies. But, ironically, "too many of these initiatives simply paved the cow paths" of outdated technology (Tapscott, quoted in Lathrop and Ruma 2010: xvi). In other words, they focused on automating existing processes—which created efficiencies but did not necessarily enhance effectiveness.

By contrast, a next wave of innovation presents a historic occasion to fundamentally redesign how government operates, how and what the public sector provides, and ultimately how governments interact and engage with their citizens. As Anthony Williams and Heidi Hay observed over 10 years ago, "the core of digital-era policymaking is citizen-centric process that requires active and informed participation by citizens themselves. … (A)uthority and legitimacy comes from the citizenry" (Williams and Hay 2000: 4). Writing more recently on behalf of the Lisbon Council, Williams says (2010a: 26):

Digital citizens increasingly expect to be partners in governance, not bystanders. It is time governments at all levels abandon their monopoly over the policy process in favor of participatory models that invite input—and ownership—at all stages of development, from problem definition and analysis, to identifying strategic options and making decisions. This goes far beyond the occasional Internet consultations that for instance the European Commission conducts, or blog of a government official. Instead, it is a process of opening up processes that have hitherto been closed and making governance and government more transparent, more accountable and more understandable.

The end result of such change would be a "massive power shift from governments to citizens, as we no longer have to rely on bureaucracies to dictate what information we need and what we must do with it" (Eggers 2007: 16).

Fortunately, there is evidence that a number of governments in North America, Europe and elsewhere have embraced "citizen-centric" approaches to service delivery and begun to emphasize interagency collaboration. Such approaches demonstrate the "reality ... that citizens can self-organize to do many of the things that governments do today, only they often do them better" (Williams 2010a: 21–22). What form of self-organization does the Web enable? Williams believes that one analog to the ecosystems in the business world that are enabled by virtual platforms are "policy webs," which are "emerging as the leading organizational form for enabling greater innovation, agility and citizen participation in policy-making" (Williams and Hay 2000: 3). Policy webs are "Internet-enabled networks of participants that contribute a broad range of skills, experiences, perspectives and resources to constitute an effective policy-making ... (and that draw) participants widely from governments, international organizations, businesses and industry associations, think tanks, academic institutions, civil society organizations (e.g., NGOs, associations and religious groups), and the general public" (ibid.).

More recently, Williams and Don Tapscott observed that "in the new model of public service delivery, the 'citizen collaborator' becomes a prosumer of services"—that is, both producer and consumer—"identifying needs and helping to shape their fulfillment" (Tapscott and Williams 2010: 268). In the section that follows, we will provide examples of such collaboration from several countries, as well as examples of Web-enhanced participatory budgeting, which can now be found in places such as Brazil, China and Germany (ibid.: 269). Moreover, we will summarize several visions for forms of Gov 2.0 in which external experts and passionate citizens augment the bureaucracy's monopoly on "policy expertise."

Realizing the full potential of democracy

Although the public sector has not been a pacesetter in pioneering Web 2.0 activities, there is reason to believe that it could blaze revolutionary trails. The collection of essays in the excellent anthology *Open Government* (Lathrop and Ruma 2010) paints an inspirational picture of what might lie before us. Here are a few of the possibilities envisioned by those and other advocates of Gov 2.0:

- *Collaborative democracy:* One example of this would be using wikis to move beyond the limited concept—and even more limited accomplishments—of "deliberative democracy," which stresses equal representation in participation and enshrines consensus as an end in itself, to "collaborative democracy," which instead focuses on matching participation to interest and skills and sees good decisions as the end. Using a pilot experiment in the *U.S. Patent and Trademark Office* as a case example, Beth Simone Noveck argues that this is quite feasible (2009). It is an encouraging sign that, as of this writing, Noveck is on leave from her academic roles,[81] having been invited to join the Obama administration as U.S. Deputy Chief Technology Officer for Open Government, where she directs the White House's *Open Government Initiative.*

- *The "long tail" of public policy:* David Eaves argues that truly open government is becoming more possible because the Web enables a public policy equivalent of the "long tail" that Christopher Anderson has described (Anderson 2008). In other words, in the same way that the economies of the Web create profitable niches in markets so small and/or specialized that none was possible before (as has been proven, for example, by *Amazon.com*), the Web also enables "a long tail of interest and expertise which, thanks to collapsing transaction costs," can now self-organize (in Lathrop and Ruma 2010: 191–193). This phenomenon could in fact enable Noveck's "collaborative democracy."

- *Self-designed personalized government:* "By arming people with useful information about quality, cost and performance, government can … adopt a less paternalistic approach to their 'customer' and shift many programs from monopoly to choice-based models. … One-size-fits-all government—designed by politicians and bureaucracies for the ease of politicians and bureaucrats—can be transformed into 'government you design,' where everything we want to do involving government can be customized to each individual's interest, location and needs" (Eggers 2007: 16). Taken to its logical extreme, this could result in "opening up free trade in government services," in which citizens could seek out their own solutions from a global cross-sectoral community rather than a single entity (Tapscott and Williams 2010).

- *Informal government on a large scale:* Advocates of "emergent democracy" foresee a future in which the tools of the Web enable citizens to interact in ways that capture on a formal basis and a large scale the virtues of informal self-government (which was possible up until recently on a community level and involved no more than about 150 people). As Armstrong writes: "It's possible to conceive of large-scale democratic systems an order of magnitude more complex than existing ones that harness our complex social behavior for collective decision-making rather than disabling it" (Armstrong 2010: 171). In a world that is deeply Web-enabled, complex and powerful forms of participative democracy (e.g., the Themis Constitution, in which "each citizen has an equal right to propose and vote on group decisions") become feasible on a large scale for the first time. An early and encouraging experiment was the *Chaos Communications Congress,* the world's first "virtual organization" (ibid.: 171–174). The participants in this virtual community were all highly tech-savvy. Although creating ways for the average citizen to engage in this way would be challenging, the potential is still there.

The above possibilities illustrate that there is still a long way to go before the full potential of Gov 2.0 is realized, even in the countries that have pioneered the most advanced applications. At the same time, the examples that follow will show that much has been accomplished and in surprisingly diverse parts of the world.

Patterns and Examples at the International Level

- In 2008, the European Commission launched its *Open Access Pilot for Research Project*. The project made available €50 billion for publicly funded research enabling the Internet to go beyond dissemination of scientific information to become the new platform for doing science and conducting research (Williams 2010a: 7). Indeed, the average number of coauthors writing a paper has more than doubled in recent years, and a growing number of scientific papers have between 200 to 500 coauthors, with one study having an astonishing 1,681 joint authors (ibid.).
- The European Environment Agency's "Eye on Earth" portal offers an interactive map from which citizens can get real-time access to data on the quality of air and bathing water from the 32 member countries of the EEA. Users can browse the visual imaging interfaces and drill down for detailed, neighborhood-level data about ozone levels, nitrogen dioxide, particulate matter and carbon emissions. The site also features social-networking applications and discussion forums in which citizens can debate what the data mean or help raise awareness of environmental problems in their own communities. Citizens can even contribute their own data and observations about the environment around them, including first-hand experiences of climate change and potential explanations for environmental degradation in specific areas (Williams 2009: 24).

Patterns and Examples at the National/Federal Level

A handful of national governments throughout the world have made significant commitments to Gov 2.0. This section provides brief profiles of a sampling of national initiatives, concluding with a much more comprehensive profile of the range of activities visible at the federal level of the government of the United States.

- *Australia:* In May 2010, Australia's federal government accepted 11 out of 12 recommendations from its "Government 2.0 Taskforce." The task force's central recommendation was that the government make a declaration of open government. The government accepted this recommendation and announced that it expected to make such a declaration in the following months.[82] Still, a number of examples of open government were initiated years ago. For example, the *Australian Tax Office's*

"Listening to the Community" program aims to create a more user-friendly tax system by involving stakeholders at every stage of the design process. Citizen input obtained through field visits, focus groups, prototype development and product-testing is being used to continually refine this system. The agency has even created a simulation center where users and co-designers work together to troubleshoot problems and test products (Eggers 2007: 248). Likewise, Australia's "Centrelink" agency is one of the world's largest seamless government-integration projects. The agency has assembled a wide variety of social services (ranging from social security to employment) from 13 federal departments, various state and territorial governments, and nonprofit providers, and gathered them all together to allow "one-stop shopping." The program offers a combination of office-based and Internet services that produce around 3 billion transactions a year. "Our whole thesis," says CIO Jane Treadwell, "is that there's value in having a national architecture but the major ingredient needs to be local context, local players. We call it 'linking up'" (ibid.: 51).

- *Austria:* Austria has become an e-government role model within Europe (Gegenhuber 2010). All levels of the government (federal, provincial and local) and key stakeholders offer important services that enable citizens to cooperate on the "digital Austria" platform. The site description reads (in English):

  > The platform digital Austria is the coordination and strategy committee of the Federal Government for eGovernment in Austria. eGovernment includes the totality of all electronic public administration services for the Austrian people. With it, the access to and the contact with public authorities become easier. More than 80 percent of the enterprises already use eGovernment services, more and more citizens are electronic customers. eGovernment is a synonym for a modern and innovative state in which quality, trust and speed are central elements. In the sense of a one-stop-shop, the platform *digital Austria* offers comprehensive information about eGovernment on the following sites. This principle is extended on as many areas as possible:
  > - To be able to do inquiries or file an application electronically;
  > - To be able to receive information electronically at any place in Austria;
  > - To ease the handling of administrative procedures.

- *Germany:* The *Open Data Network Deutschland* offers an online catalogue of comprehensive databases of information on public government, libraries, science, research and politics.
- *Singapore:* The *eCitizen* website, which offers more than 500 online services, was one of the first public portals organized around "life events" (e.g., birth registries, primary and secondary school services, job-search and career services, housing aid and retirement services) rather than by department. The site has sections called "towns" dealing with elections, libraries, sports, recreation and travel (ibid.: 18).
- *India:* "The government of India has placed kiosks connected to the Internet in hundreds of small towns so that rural farmers (can) get information about crop

markets. The information helps farmers negotiate better prices with middlemen buyers" (ibid.: 20).

- *China:* When a prisoner escapes from a prison in a populated area, wardens are alerted by text message and then images of the escapee are distributed by multimedia messaging to local residents so as not to scare them with a siren (ibid.: 28).

Comprehensive Case: The United States

As Eggers writes (Eggers 2007: 245): "In 2006, the U.S. government did what no country in the world had ever done. By launching Expectmore.gov, it opened a window into the performance of the federal government." Remarkably, it also "opened up its performance—warts and all—to unprecedented scrutiny to an unprecedented extent" (Dixon et al. 2005: 246). Most surprising is that this happened under the presidency of George W. Bush, who has been widely accused of being among the most secretive chief executives in American history. According to its creator, Budget Deputy Director Clay Johnson (who was also President Bush's "best friend"), the website was to some extent a reaction to just this criticism. As Johnson put it: "Public shaming sometimes works" (ibid.). In any case, the U.S. federal government has gone on to make the most significant commitment to Government 2.0 of any nation.

The symbolic shift of the Bush administration in its waning days was followed by overt advocacy by the incoming Obama administration. Immediately upon his inauguration, President Obama signed a memo entitled "Transparency and Open Government."[83] In May 2009, the White House's *Office of Management and Budget* (OMB) followed up by releasing a landmark directive that detailed how federal agencies will adopt innovative online tools (e.g., *Twitter*) and utilize social media sites to enhance interagency collaboration, increase transparency and foster citizen participation in agency decision-making.

The new administration found that the U.S. government had been outpaced by other sectors by a staggering degree. This came as a rude shock to a tech-savvy member of the Obama transition team tasked with prepping the new administration for inauguration day. Although he tried to "bring Web 2.0" to Washington, he ran into a number of formidable blocks. A short list of these barriers includes:

- Restriction on services carrying ads, which many Web services do;
- A prohibition against unlimited liability clauses, which most Web services have;
- Laws requiring equal access for the handicapped, which dictate that videos on sites such as *YouTube* be transcribed; and
- The Presidential Records Act, which requires that all documentary materials related to the presidential office must be saved for posterity, thereby theoretically requiring everything accessed on the Web to be printed out.[84]

More disturbingly, the intelligence agencies had continued to lag far behind in taking advantage of Web 2.0. Some observers had speculated that the 9/11 attacks might have been prevented if the intelligence communities had used these technologies to help "connect the dots" (Thompson 2006). In any case, the adherence to classic bureaucratic forms of organization in those agencies contrasts sharply with the flexible strategies employed by terrorist groups such as al-Qaida, which arise from their capacity to flexibly adapt through spontaneous self-organization. For such loosely organized networks, having their elusive leader, Osama bin Laden, captured would probably have more symbolic value than substantive impact. To be sure, due to their need for security, the intelligence agencies have a particular challenge when it comes to openness.

Adding new initiatives to those that preceded the Obama administration, the U.S. government can now point to a wide range of robust practices. While it still falls far short of exploiting the Web's full potential, it has clearly demonstrated the value of using the tools of Gov 2.0 in pursuit of the goals of open government.

White House initiatives

- The new administration announced a website—www.recovery.gov—that enables the public to monitor implementation of the economic stimulus package. The administration also issued an invitation to submit ideas for making the website a "more effective portal for transparency."[85]
- Data.gov, launched in 2009, enables anyone to search and download federal data sets.
- In April 2010, President Obama announced the release of open government plans by all cabinet agencies (Noveck 2010). For example:
  - The *U.S. Department of Housing and Urban Development* (HUD) is recording all of its public events and making them available online.
  - The *U.S. Department of Education* is publishing its secretary's schedule.
  - The *U.S. Department of Labor* announced the release of its new "Online Enforcement Database," making all workplace-safety data searchable in one place.
  - The White House's chief technology officer and chief information officer published a "dashboard" ("Around the Government") that reports progress on these issues.
- In mid-2009, the *General Services Administration* reached an agreement with *Facebook* and *MySpace* to resolve legal concerns that had been a barrier to government organizations that wanted to use the sites (Tartakoff 2009). Vivek Kundra, the first chief information officer (CIO) of the United States, is encouraging agencies in the U.S. government to use free Google services and open source wikis. As Tapscott and Williams (2010) write: "(Kundra) calls it the government cloud, but think 'apps store for government'—a place where employees can access a vast ecosystem of secure applications and data sets for doing their jobs." (For a description of what

Kundra accomplished during his tenure as CIO of the District of Columbia, please see below.)

Coordination and communication among federal departments

- Intellipedia,[86] an online system for collaborative data-sharing among the 16 agencies in the U.S. intelligence community, was launched in 2005 (Wikipedia; Evans 2009; Jackson 2009).
- The *U.S. Department of Health and Human Services* (HHS) will create a new Web-supported interagency task force to improve and coordinate how the government implements health information technology.[87]

Initiatives by federal departments

In May 2010, the *Recovery Accountability and Transparency Board* announced it would be the first government-wide system to be moving Recovery.gov to the cloud. In April 2010, the *Department of Health and Human Services* began using the cloud to support implementation of the *Electronic Health Records* (EHR) systems.

- *CompanyCommand:* An example of a public-sector Web 2.0 application that has had a major impact is the *U.S. Army's* "CompanyCommand," which connects company commanders—past, present and future—so they can share knowledge (Dixon et al. 2005). Army leaders are able to obtain support in dealing with whatever challenges may emerge, such as how to communicate with family members of soldiers killed in action. CompanyCommand has evolved from a base consisting of an interactive website providing cutting-edge, world-class resources. It has generated a book, a monthly newsletter and face-to-face learning opportunities. One particularly dramatic example of connecting leaders *in* the experience to leaders *with* experience occurred when the idea evolved of setting up workshops led by combat-experienced officers returning from Iraq with company leaders about to be deployed. Such "just-in-time" learning enabled those about to lead men under life-threatening conditions to benefit from the most recent lessons learned by their peers who had faced similar challenges (ibid.: 23–31). In so doing, CompanyCommand serves as a vehicle for mentoring. A study of the community concludes that "connecting leaders in conversation about their work transforms the individuals who participate as well as the whole profession" (ibid.: 178).
- *NASA's Clickworkers*[88] invited public volunteers to identify craters on Mars from images available online from the space probes *Viking Orbiter* and *Mars Global Surveyor* (Monitor Institute and David and Lucile Packard Foundation 2007: 9).
- The website of the *Mental Health Services Administration* of HHS has an online facility locator for substance abuse treatment that allows family members, social

service workers or the substance abusers themselves to search for treatment centers in any part of the country. Based on answers to a series of questions, the site will suggest several different facilities that meet the characteristics the individual or family member has indicated are important to them (Eggers 2007: 22).

- In the *U.S. Coast Guard*, security sensors are being used to track ships and containers. They send information to central authorities about where the ships have been and whether any containers have been breached (ibid.: 28).

Ecosystem of communities of practice

- *GovLoop* is an open online platform that allows individuals and organizations to develop their own hubs for social networking. Founded in 2008, it already has over 30,000 members made up of U.S. government employees, academics, students and members of interested social-sector organizations. It has also attracted civil servants from Brazil, Canada, Israel and the United Kingdom. It features newsfeeds, forums, job and event boards, blogs and profile pages, but it costs only $100/month to maintain. GovLoop was founded by Steve Ressler, an employee frustrated by the Department of Homeland Security's ban on Facebook. He views 30,000 as "just the tip of the iceberg" and adds that (quoted in Tapscott and Williams 2010: 277): "People really want to learn from what others are doing. If we can strengthen engagement, foster more dialogue and connect more people, we can turn conversation into knowledge."
- *GovLeaders.org* enables access to leadership resources and features blogs on public-sector leadership.
- A *Government 2.0 Club* was established in 2009 to "bring together thought leaders in government, academia and industry from across the country to explore how social media and Web 2.0 technologies can create a more transparent, participatory and collaborative government" (Drapeau 2009).

Comprehensive case of a federal department

- *The U.S. Department of State:* The State Department has made headlines under Secretary of State Hillary Clinton by using social media to reach out to a wider audience (Garekar 2010). Her department was able to build on an initiative set up under the Bush administration that developed an official blog called *DipNote*. As of this writing, the blog has had 12 million page views, thanks in part to its new additions, including a *YouTube* channel, *Twitter* feed, *Facebook* page and *Flickr* photo account. The State Department has also developed a "Social Media Field Guide" for Facebook pages so that embassies can create their own pages in compliance with policy and legal constraints. For example, the webpage for the U.S. Embassy in Jakarta is written in the local language and has almost 20,000 fans. In 2009, Secretary Clinton announced the establishment of "Civil Society 2.0" to "help grassroots organizations around the world use digital technology to tell their stories, build their memberships and support bases, and connect to their community of peers around the world."

By some accounts, the State Department is "far ahead of many organizations in terms of being open and using social technologies—even though, being part of a government agency, staff face many restriction and have far more at risk in terms of international engagement" (Li 2010: 266). This can be attributed to cultural factors, which underscores the importance of that variable. Individuals "already had tremendous freedom to act independently in accordance with the circumstances they face on the ground. They are trusted to develop the relationships they think are best suited to achieving their diplomatic goals" (ibid.) The interaction with the culture proved to be mutually advantageous. Diplomats were able to extend the diplomatic culture to the online space, making it easier for them to "get over their fear of being open" (ibid.). The net result has been a transformation of the "overall relationship between the organization and its constituents" (ibid.: 264)

The State Department also had an Office of *eDiplomacy* as early as 2003. Its mission is to promote end-user involvement in decision-making about information technology, to improve the way the department works with its foreign-affairs partner agencies and other entities, and to foster knowledge management within the entire department. The related *Diplopedia* provides a living repository of organizational knowledge and enables officials within the U.S. Foreign Service to share vital info with colleagues around the world.

This case and many other examples in the U.S. federal government illustrate that, in the words of one State Department employee, the 21st century "is really a terrible time to be a control freak" in the government, as well (Lichtenstein 2010).

Patterns and Examples at the Regional/State/Local Level

For years, open government primarily meant enacting sunshine laws to guarantee access to data held by the state. But online technology has helped expand the definition of open government, enabling states and municipalities to tap into citizen ingenuity. Indeed, state and local governments now provide citizens with unprecedented access to their services while simultaneously streamlining their processes and increasing efficiency.

Enabling more representative democracy

- Roughly 2,500 years ago, the Athenian democracy employed the "kleroterion," a device for random selection, as a means of engaging citizens in ways that balance representation with efficiency. The device picked 500 people at random to make policy for the city-state's 50,000 citizens. Today, a Web "mashup" links modern technology with this age-old practice as a new form of "deliberative democracy." Professor James Fishkin of *Stanford University* adds digital tools and other techniques to this practice. He starts by canvassing views from a large sample of people. Then, a smaller subset (normally around 300) receives briefing materials from the opposing advocacy groups. Moderators lead small-group discussions that subsequently draw up questions for experts and policymakers. Statistically representative samples aim to give a credible picture of what the entire population would think if it were as well-informed. Such polls have covered aboriginal rights in Australia and traffic jams in La Plata, Argentina. A Chinese jury is using this approach to look at land rights. In Fujisawa, Japan, a poll on local-government planning started in August 2010 ("Deliberative Democracy" 2010).
- The *Institute for 21st Century Agoras* is a nonprofit organization "dedicated to cultivating authentic democracy within the Information Age." In projects such as the *Cyprus Civil Society Dialogue*, it creates platforms for "civil dialogue for high-complexity problems in face-to-face, virtual and mixed participation engagements."

Making government transparent

- South Dakota's *Department of Public Safety* began sending out text messages in March 2009 to alert motorists of upcoming sobriety checkpoints as a way of deterring drunken driving (Martin 2010).
- In 2010, prodded by Web-savvy state employees in the administration of Governor Deval Patrick, the *Massachusetts Bay Transit Authority* shifted from a bunker mentality toward its information to a posture of sharing it openly to make real-time scheduling information available to passengers (Govmonitor 2010).

Bringing the people into government

- Anthony Williams of the *Lisbon Council* writes (Williams 2009: 24): "Europe is flush with ... examples of citizens taking on larger roles in their communities and demanding greater ownership of their government and democracy. In the Netherlands, the cities of Apeldoorn, Helmond and Tilburg are using virtual worlds to engage citizens in the planning and development of local development projects. Interested citizens can learn about recent developments and voice their opinions on issues ranging from aesthetics to traffic congestion and pedestrian safety at vir-tueelNL.nl. In the case of Tilburg, they can vote directly on a selection of virtual mock-ups for a new central marketplace in their city."

- Stimulated by "Apps for Democracy," many American cities (e.g., Portland, Oregon) and states (e.g., New York) have started competitions offering prize money to developers who build software applications using public data. One winner of the New York competition, *MyCityWay*, makes a smartphone app that helps users find restrooms, Wi-Fi hotspots, subway stations and more.
- In Nova Scotia, the "Imagine Halifax" project created a citizens' forum for elections in 2004. It brought together several dozen activists to compile a platform using live meeting and e-mail with follow-up from *seedwiki*.[89]
- The *Green Party of Canada's Living Platform* built on this precedent in 2004–2005 to launch a more thoroughly designed initiative, compiling citizen, member and expert opinions in co-developing its platform. During the election, it gathered both input from "Internet trolls" and the support of other parties. Ironically, the initiative was ultimately derailed by Jim Harris, the party's leader, when he discovered that it was a threat to his status as a party boss. The site openpolitics.ca eventually evolved out of this effort.[90]

Facilitating communication of citizen concerns

- The city of New Haven, Connecticut, used *Google Maps* to help people flag issues in their neighborhoods and send notices to their local officials. According to Tozzi (2010), 36,000 problems have been reported, and 40 percent of them solved.
- *SeeClickFix*, a for-profit enterprise founded in the United Kingdom, enables citizens to report problems (e.g., potholes and graffiti) to the city government. Usage has since spread to the United States and other countries.
- *FixMyStreet.com* enables residents to directly submit concerns about safety, vandalism or other local issues to their municipal council. As Williams writes: "Set up by a nonprofit called *mySociety*, the site is part of a larger trend as public agencies across Europe build online innovation spaces where the general public and staff can co-create information-based public services, much the way companies such as *Amazon.com*, *Flickr* and *Apple* enable third-party developers to build extensions of their software platforms" (Williams 2010a: 22).
- San Francisco, California, is the first city to take advantage of the *Open 311 API* jointly announced in February 2010 by the city's mayor, Gavin Newson, and U.S. Government CIO Vivek Kundra. Almost every city in the United States has a "311" customer service center that gives residents answers about city services and provides a place to report potholes, graffiti and other issues. Between March 2007 and the time of this writing, San Francisco's 311 had logged more than 8 million calls.

Later in 2007, the city took a step toward Gov 2.0 by initiating a partnership with Twitter. The Open 311 API standard allows software developers to write Web applications that obtain service request data from the 311 system and submit new service requests to city departments. San Francisco collaborated with *SeeClickFix* to develop its own program.

- In Barcelona, Spain, residents use the IRIS system to call a toll-free phone number to access services or register a complaint. Once a request is resolved, callers receive a personalized reply through their choice of SMS, e-mail or regular mail (Eggers 2007: 249).

Cases at the Regional/State/Local Level

- *Rubbish collection in Estonia:* Anthony Williams has documented a novel example of how a combination of Web-based tools enabled Estonians to deal with a very unusual problem (Williams 2008b):

  When Estonians regained independence from the former Soviet Union in 1991, they not only acquired new political freedoms, they inherited a mass of rubbish— thousands and thousands of tons of it scattered across illegal dumping sites around the country. When concerned citizens decided that the time had come to clean it up, they turned not to the government, but to tens of thousands of their peers. Using a combination of global positioning systems and *Google Maps*, two entrepreneurs (*Skype* guru Ahti Heinla and *Microlink* and *Delfi* founder Rainer Nolvak) enlisted volunteers to plot the locations of over 10,000 illegal dump sites, including detailed descriptions and photos. That, in itself, was ambitious. Phase II of the clean-up initiative was, by their own admission, rather outrageous: clean upwards of 80 percent of the illegal sites in one day using mass collaboration. So, on May 3, 2008, more than 50,000 people scoured fields, streets, forests and riverbanks across the country, picking up everything from tractor batteries to paint tins. Much of this junk was ferried to central dumps, often in the vehicles of volunteers.

- *Neighborhood Knowledge California (NKCA):* Anthony Williams has also provided a profile of an innovative use of new Web tools to identify troubled communities (Williams 2008a):

  NKCA knits together municipal databases and inspection records, looks for indicators of urban decay and plots the information on local and state-level maps posted online. Rather than having to look at each database separately, public officials, citizens and businesses can search by zip code or other parameters to view comprehensive information on one property or see at a glance which communities might be headed for trouble. Private-sector developers can spot potential investment opportunities (e.g., a cluster of buildings in financial difficulty), while community organizations are using the NKCA as a tool for community empowerment. NKCA even has a code-enforcement tracking system that lets residents monitor the City of Los Angeles' responses to housing code complaints. ... Data that might otherwise have gone unused in filing cabinets is suddenly a catalyst for better policy-making, more effective local government and neighborhood economic development.

- *Linz, Austria:* Linz, a city of just under 200,000 (with another 70,000 in the greater metropolitan area) is setting the pace in Europe. Linz native Thomas Gegenhuber, a researcher for nGenera, described with enthusiasm three initiatives that he believes illustrate the kinds of innovations the city is piloting:[91]
  - *Creative Commons licensing of publicly subsidized work:* Linz publishes everything using a *Creative Commons* (CC) license (noncommercial version) approach that emphasizes open data. The rationale is that since the taxpayer pays for data, it makes no sense for it to be proprietary. Artists who receive subsidies in Linz get 10 percent more if they make their work available via CC.
  - *Availability of a public-space server:* Linz provides digital "public space," which is the Internet equivalent of constitutional guarantees of fundamental rights (e.g., freedom of speech, the right to demonstrate, the right to form political parties, etc.).[92]
  - *Free public Wi-Fi:* Linz has 120 squares with free Wi-Fi. The city is not unique in this regard, but it is unusual. After all, most airports are still struggling to provide such access.
- *Freiburg, Germany:*[93] As mentioned in the narrative, "participatory budgeting"— that is, the direct involvement of citizens in public spending decisions—began in Brazil but is spreading across the globe and being enhanced by online tools. Examples include the small town of Jun in Granada, Spain (where citizens can participate live during the budget debate using mobile phones and wireless Internet), the city of Parma in Italy and the French region of Poitou-Charentes (Jellinek 2008). Germany has at least two examples, the best known of which is the city-state of Hamburg, which used software that has since been made available by its creator, *Demos Budget*, as a generic system. The southern German city of Freiburg is one of the municipalities that has used the software. We chose to profile Freiburg's more recent initiative because it is a less well-known example in a more typical setting.

In his description of an "innovative case of public policy consultation," Alex Marshall (2009) writes:

In 2008, the municipal government of Freiburg invited its citizens to take part in a participatory budgeting exercise. The goal was to gather citizen input for the drafting of the 2009/2010 municipal budget. With the help of software company TuTech Innovation (which created the software used in Hamburg), the Freiburg government created a website that used discussion forums, wikis and a new innovation—the budget slider.

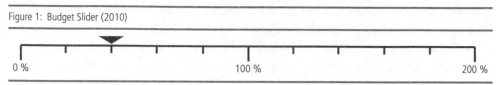

Figure 1: Budget Slider (2010)

Source: Marshall 2009

Citizens who registered for the website could manipulate these sliders to create their own individual budgets by moving the sliders up or down to either increase or decrease spending in any one of the 22 budget areas. The key constraint was that the total budget had to balance to 2008 levels, so spending increases in one area necessitated economizations in another. Citizens were also invited to provide written justifications for their changes. Following the completion of the process, all of the individual budgets were aggregated into one single "Citizen's Budget," which gave a clear picture of the participants' wishes for the 2009/2010 municipal budget.

Overall, 1,800 citizens registered to use the website, with 1,291 writing individual budgets (750 of whom provided written justifications). Although this is less than 1 percent of the city's population (217,000), it still represents a sevenfold increase over the roughly 150–200 citizens who might show up for an offline, town-hall consultation process. Building on the Change.gov model, this input was actually used as a focal point in the local government's debate over the drafting of the actual budget. In one case, €400,000 were redirected to childcare spending, a change that may not have occurred without the widespread support that the measure received in the Citizen's Budget. Also building on the suggestion box model, the final Citizen's Budget was drafted into a report that was published by the municipal government. This allowed a great deal of transparency, as this budget could now be compared to the actual budget that was written into law, also providing an improved degree of government accountability. Overall, this case demonstrates the new relationship that's possible between government and citizens. Simple tools like the budget slider can add a whole new level of transparency to the public square dialogue.

Critics of the Freiburg example note that the outcome of the participatory budgeting (i.e., more resources for child care) was no doubt related to the composition of the group that self-selected to participate—a group whose members are more educated, make more use of the Web and more are savvy about it. Such self-selection clearly introduces a bias in the process.[94] More representative outcomes will require higher levels of citizen Web engagement.

- *Washington, D.C.:* The District of Columbia became a radical pioneer in Gov 2.0 under the leadership of CIO Vivek Kundra, who was then recruited by the Obama administration to become the first chief information officer of the United States. Kundra migrated the District of Columbia's IT infrastructure from organizationally based "enterprise" platforms to Web-based solutions (e.g., Google Apps), thereby achieving a 90 percent reduction in costs. He also created a city-wide data warehouse enabling all government employees and stakeholders to see and help analyze what does not work in the community. Most visibly, he put up $20,000 in prize money for an innovation contest, "Apps for Democracy," which resulted in 47 Web, iPhone and Facebook apps in 30 days, delivering an estimated $2.3 million in value to the city for a total cost of $50,000 (Tapscott and Williams 2010).

- *Links to further city-wide cases that are often cited:*
Manor, Texas, www.manorlabs.org/
Edmonton, Toronto and Vancouver, Canada

How the Web is Fostering the Emergence of a 21st Century Commons

As we have noted throughout this report, one of the most significant societal trends with implications for leadership is the increasing complexity of problems confronting leaders in every sector (Scharmer 2009: 81–104). We even heard it from within our sponsoring organization: For example, a senior project manager in the Future Challenges program at the *Bertelsmann Stiftung* summed up his experience: "Our conclusion was that complexity has always been around. But due to the increasing number of global interfaces and the networking opportunities offered by the Internet, what we are now witnessing is exponential growth in the level and dynamism of global contexts."[95]

Recognition that problems cannot be solved by any single organization or even multiple organizations within a single sector is becoming commonplace. The Web has facilitated this recognition by making information about social problems more readily available. Furthermore, it provides new tools for the creation of—and ongoing collaboration among—new communities. Increasingly, individuals and organizations are using the tools of the Web to come together across sectoral boundaries, finding common cause in the effort to address "wicked" problems that defy solution from within any single sector.

In reviewing the initiatives in other sectors, we began to appreciate a range of activities that could not be easily understood within the boundaries of the three traditional sectors. We have lumped them together as the global expression of an old institution, the "commons," which is a central gathering place in a community where people come together to solve problems and celebrate all that binds them together.

We see this new "sector" (we will call it that for our present purposes) as a vital source of new leadership for addressing intransigent global problems. We also see this sector—or at least its constituent activities—as continuing to become more and more significant. A report sponsored by the James Irvine Foundation found that "sector boundaries are blurring" (La Piana Consulting 2009: 16). We envision the possibility that the "Global Commons" will eventually subsume in large measure the other, more discrete sectors as people within, across or outside organizations rise to the challenge of collaboratively constructing sustainable lifestyles, cultures and societies in a world of increasing complexity, accelerating change and daunting problems.[96]

Literally on the final day of drafting this report, we were delighted to discover that our notion of a "commons" constituting a new sector can build upon the "reinvention of the commons" proposed by Peter Barnes (2006). Barnes provides considerably more depth regarding historical background and current examples, though with no reference to the Web. Our belief is that the commons he advocates is already constructing itself thanks to the Web. We include five constituent elements in this "sector," each of which has been—or soon will be—significantly shaped by the tools of the evolving Web.

Social Entrepreneurship

Social entrepreneurship (leadership) comes from individuals and small groups that see an opportunity in an unmet need. Such initiatives serve the purposes of the social sector while using the means of the business sector. The term has a long history (Bornstein 2007) and has been traditionally regarded as an innovation within the social sector (or, to be more precise, the "citizen sector," as it was dubbed by William Drayton, who founded Ashoka, the organization that is perhaps the most prominent example and sponsor of social entrepreneurship). However, we see the application of business-sector methods (sometimes by for-profit entities) to achieve social-sector goals as a reason for including social entrepreneurship among the foundational elements/components of the emerging Commons.

Organizations acting as social entrepreneurs have demonstrated the ability to push businesses and consumers to expect new standards in the marketplace by showing that products can be produced using fair-trade policies or with less environmental impact (Bernholz, Skloot and Valera 2010). However, they are at an inherent competitive disadvantage once traditional businesses follow their lead into the marketplace. For this reason, they are unlikely to achieve maximum impact through financial scale. Instead, their power lies in using networks for a variety of purposes: to share technology, to jointly produce goods and services, and to purchase inputs as a group. Web technologies form a critical infrastructure for such networks.

Today's social entrepreneurs straddle the boundary between the business and non-profit sectors. As Bernholz et al. write (ibid.: 39):

> On the one hand, we see a proliferation of phenomena that harness market mechanisms to solve social problems: socially responsible investing, information marketplaces such as the FasterCures Philanthropy Advisory Service, B Corporations, low-profit limited liability companies (L3Cs). On the other hand, we see an enormous commitment of time, energy, ingenuity and know-how to nonmarket, nonproprietary phenomena that are themselves social goods: open source software, wikis, Project Gutenberg.

In making sense of this trend, Bernholz et al. point to the "blended value" proposition developed by Jed Emerson, which states that "all organizations, whether for-profit or not, create value that consists of economic, social and environmental value components—and that investors (whether market-rate, charitable or some mix of the two) simultaneously generate all three forms of value through providing capital to organizations" (quoted in Bernholz, Skloot and Varela 2010: 38). The study's authors conclude (ibid.: 39): "We may be approaching a moment when the idea of blended value, which resolves the contradiction between market and nonmarket impulses, may become as commonplace as belief in the 'invisible hand' of the market is today." Indeed, Lucy Bernholz predicts that (ibid.: 38): "We will not only see the blending of market and nonmarket organizations, we will see the corresponding development of new approaches to funding, finance and reporting requirements."

Following are some examples of social entrepreneurship involving cross-sector initiatives:

- Based in London, *SIX* (Social Innovation Exchange) is a global community of over 700 individuals and organizations (including small NGOs and global firms, public agencies and academics) committed to promoting social innovation and growing the capacity of the field. Its website says: "SIX was designed to fill a gap." It intends to link existing organizations together "to promote learning and collaboration across sectors, fields and countries." Williams sees it as "an example of a steady rise in organizations engaging in social innovation." He observes that it is "Internet-savvy and fond of social networking and collaborative tools" (Williams 2009: 24).
- On the *Pew Charitable Trusts' Subsidyscope* website, users can track federal government subsidies.
- Estonia's mass cleanup on May 3, 2008 (previously discussed above) is an example of an effective use of Web 2.0 in support of a network. Anthony Williams reports that "networks are increasingly beating institutions at their own game" (Williams 2009: 25). Networks have become gathering points of knowledge that can concert into rapid action. He concludes that "to remain relevant, governments will need to be agile, open and fluent in the ways and means of collaboration" (ibid.).
- *Microsoft's Corporate Citizenship program*, which Beth Kanter describes as "Microsoft's networked approach to accelerating social change through technology," focuses on technology and partnerships that can accelerate systemic social change and economic issues. Areas of engagement include workforce development, employee giving, environment, online safety and the role of technology in the nonprofit community. Microsoft also has a partnership with *TechSoupGlobal*. The Corporate Citizenship program was ranked 14th on a list of the 100 best corporate citizens by *Corporate Responsibility* magazine, which evaluated performance on a range of issues such as environment, climate change, employee relations, human rights and philanthropy (Kanter 2010g).
- *FSG Social Impact Consultants* is a nonprofit organization that works with foundations, corporations, governments and other nonprofits to "accelerate the pace of

social progress by engaging civil society through offering advice, distilling and sharing learning, and creating new products and services." For example, Grant-makers for Effective Organizations sponsored a webinar on a new approach to "shared measurement" of impact, hosted by two FSG staff members who had written a book on the subject.

- The *U.S. federal government's* attention is directed to all three categories of Web users (citizens, businesses and government) as well as "customer segments" within each role. As Eggers writes (Eggers 2007: 19): "Previously, if you fell into one of these categories, you had to interact with a baker's dozen of different federal agencies."
- Eggers suggests that by "partnering with market-savvy private organizations," government can increase electronic government (eGov) participation. So-called channel partnerships can be established with banks and brokerage firms, sporting goods stores, trade associations and other companies or organizations that have gained the trust of customers whom the government wants to reach (ibid.: 31).
- Hans Schoenberg and Brandon Jackson cofounded *GiftFlow*,[97] a platform that "connects community organizations, businesses, governments and neighbors into a giant network of reciprocity, where they can share resources, meet each others' needs and coordinate their efforts to build a better world." According to the website, it is "a community where you can get involved by getting things you need for free and finding people who need the stuff you have to give away."

Cross-Sectoral Communities of Practice

We are learning that communities of practice are a critical resource for developing leadership capacity (Wenger and Snyder 2006). Since many of these are within organizations, they are not covered here. However, some are (primarily) bi-sectoral:
- *Leadership Learning Community's Leadership for a New Era* program has a website that fosters reflection on collective leadership and strives to build a community of practice around finding out what works in fostering collective leadership while drawing on expertise from all sectors. In 2010, it effectively used social media and a webinar to promote its initiative on leadership and race.

Others attract individuals and organizations from all sectors:
- The *Presencing Institute*, home of a community of practice growing up around Otto Scharmer's Theory U (2009), makes excellent use of a variety of Web-enabled tools (webcasts, blogs, wikis) to help fulfill its aspiration to create an ecosystem that will cultivate the leadership needed to solve intransigent global problems.
- Remarkably, there is even a purely virtual multi-sectoral community of practice, *CPSquare*, consisting of people interested in communities of practice!

Free-Agent Initiatives

In reviewing the impact of the Web in other sectors, we have repeatedly seen a shift of power away from institutions and toward networks and individuals. A symbol of this new power is the "free agent," defined earlier as an "individual working outside of organizations to organize, mobilize, raise funds and communicate with constituents" (Kanter and Fine 2010: 15). The dramatically increased ability of individuals to seize the initiative and exercise influence is a pattern that deserves to be designated as a category of activity in its own right, which promises to be one of the critical social themes of the 21st century. As the limits of leadership in established organizations in all sectors become more and more evident, free agents can be seen as a critical adaptive mechanism, serving as antennae to "sense and respond" to needs that are going unmet and problems that are either going unrecognized or are defying solutions. Thus, we need free agents to help steer us toward more just and sustainable systems. They can add momentum to the 2–3 million NGOs worldwide, helping form what Paul Hawken has described as an emerging "global immune system" (Hawken 2007: 45). Although free agents occasionally continue to work independently, Beth Kanter (2010k) finds that some go on to form their own nonprofits.

The Web is a key enabler of this phenomenon, creating possibilities that were previously only latent. As Hagel et al. write (Hagel III, Brown and Davison 2010: 67): "Using the tools and platforms emerging today, any of us can now find a person in a remote part of the world who just happens to have the knowledge or expertise required to help us out." Or, as *mySociety* founder Tom Steinberg puts it (quoted in Williams 2010a: 22): "In the past, only large companies, government or universities were able to re-use and recombine information. Now, virtually anyone with an Internet connection can mix and 'mash' data to design new ways of solving old problems." Match this potential with the "cognitive surplus" that Clay Shirky (2010) has identified in the leisure and voluntary activities of citizens, and you have the ingredients for radical social transformation. Indeed, the ingredients are already mixing themselves. Tom Watson has found that "what some refer to as online social activism and others call peer-to-peer philanthropy is quickly becoming a sector … (and) the next evolutionary phase of growth" (Watson 2009: xxi–xxii).

It is not an overstatement to call this movement "the big shift" (Hagel III, Brown and Davison 2010). Indeed, Hagel and his coauthors see "a world in which citizens gain political power relative to political institutions." It is a world in which (ibid.: 51):

- Talented employees capture economic value relative to the firm.
- Consumers have increased market power relative to vendors.
- Value is concentrated in large networks of longtime relationships rather than in transactions.
- Knowledge flows are the central opportunity, and knowledge stocks a useful by-product and key enabler.

Free-agent activity has the potential to be a catalyst in determining how leadership will be transformed in the worlds of business, government and civil society. It also offers a frame for understanding new patterns in social relationships and the pursuit of personal meaning and expression. Looking at the changes emerging at the recent turn of the century, Howard Rheingold wrote (2006): "Linux and Wikipedia ... hint at the emergence of a new information environment, one in which individuals are free to take a more active role than was possible in the industrial information economy of the twentieth century." Anthony Williams has written (Williams 2010a: 5–6): "Thanks to a whole host of new, low-cost collaborative infrastructures—such as free Internet telephony, open source software and global outsourcing—individuals and small businesses can harness world-class capabilities, access markets and serve customers in ways that only large corporations could in the past."

Yochai Benkler envisions a future that captures this potential, writing (Benkler 2006: 473): "We can make the twenty-first century one that offers individuals greater autonomy, political communities greater democracy, and societies greater opportunities for cultural self-reflection and human connection." However, he cautions that it has not yet been decided whether this will happen.

Indeed, as this report goes to press, political warfare is being waged over the ground rules that will govern the playing field and apportion power. In the United States, for example, Benkler writes that (ibid.: 385): "The battles over the institutional ecology of the digitally networked environment is (being) waged ... over how many individual users will continue to participate in making the networked information environment, and how much of the population of consumers will continue to sit on the couch and passively receive the finished goods of industrial information producers."

While the scope of influence of free agents remains questionable, there can be no doubt that their influence has already been considerable. Many of the examples cited in earlier sections are also expressions of the activity of free agents. Their activity is also evident, for example, in the open source and open software movements that have shaped the sensibility of the new generation.[98] These are examples of "commons-based peer production," which "refers to production systems that depend on individual production that is self-selected and decentralized, rather than hierarchically assigned" (ibid.: 62). This is not a new phenomenon, but "what we see in the networked information economy is a dramatic increase in the importance and the centrality of information produced in this way" (ibid.: 63).

Following are some patterns and examples of free-agent activity:

Initiating action to create or mobilize "micromovements"[99]

- One of the most remarkable and impactful examples of free agent activism is that of a hero who must remain anonymous for his personal safety. The author of "El Blog del Narco" in Mexico claims to have access to government-suppressed information about the realities and causes of the "drug wars," and the blog enables citizens to anonymously report crimes and inform on drug traffickers and other criminals. In a bizarre variation on the famous "prisoner's dilemma," drug gangs have even begun informing on each other!

- Mark Horvath turned his personal championship of the homeless into "WeAreVisible," a social-media literacy website that "helps homeless people learn how to use the Internet to tell their stories, build community and connect with support services" (Kanter 2010j).

- In late 2002, Manny Hernandez was diagnosed with type 1 diabetes. Six years later, he reflected on how he built on this personal misfortune by establishing TuDiabetes.org and thereby becoming an advocate for those who suffer from the disease. Hernandez writes that a "thought started brewing" in his head in late 2006, of making social networks "work for things beyond making friends and socializing" and putting them instead "to the service of a higher cause." He says that an article about diabetes in the *New York Times* gave him the "spark" he needed and "that Àha!' moment." He then set out to develop a social-networking site for people with diabetes. Today, as Hernandez describes it, *TuDiabetes.org* is an online community of roughly 400 people who "help each other out, educate ourselves and share the steps we take every day to stay healthy while living with this very serious condition." Hernandez also writes that the site is a place where members "write blog posts, exchange ideas in discussion forums, share photos of ourselves and our loved ones and videos that we find useful and informative" (Hernandez 2008).

- Amanda Rose launched *Twestival*, an online project for raising money for social causes, which soon became known as "Global Twestival." Not long after it was launched, Twestival took on a life of its own. Rose says: "I originally thought that we may have 50 cities involved, but only after a week of announcing it on Twitter there were over 100 cities signed up with new requests every hour. Over a dozen Twestivals were registered in the U.K. alone." She also worked closely with the nonprofit "charity: water." "Beth's Blog" also featured a profile of Amanda, capturing her reflections on what she has learned about how to be successful in undertaking this kind of initiative (Kanter 2009a).

Monitoring current events for high-impact news

- In the summer of 2008, Shekar Ramanuja Sidarth videotaped Republican Senator George Allen in a reelection campaign appearance before 100 supporters. Sidarth caught the candidate on tape referring to him as a "macaca," a racially derogatory term. Loaded onto YouTube, the video quickly "went viral" and was a critical factor in Allen's unexpected defeat (Watson 2009: 97–101). Sidarth gave interviews himself and became a "symbol of politics in the 21st century, a brave new world in which any video clip can be broadcast instantly everywhere and any 20-year-old with a camera can change the world" (Scherer 2006).

- In June 2009, anonymous protesters on the streets of Iran were able to catalyze worldwide sympathy and support by spontaneously taking a video of Neda Agha-Soltan, a protester, as she lay dying from a bullet wound on a street in Tehran. The video was uploaded to *YouTube* by cell phone.

"Liberating" information

- *Wikileaks* has generated enormous controversy by releasing classified documents about U.S. military operations and diplomatic activities in Iraq, Afghanistan and elsewhere.

Using the "nudge" approach to social change[100]

- *Carbonrally* has convinced 40,000 people to adjust their daily habits and reduce carbon dioxide emissions by some 4,500 tons. This demonstrates the power of simply making the consequences of habitual behavior apparent.

Supporting individual expression

- *Palomar5* is a network organization that aims "to create and experiment with environments that empower individuals to develop and realize positive ideas." The core project is the creation of "playgrounds" and other experiences that "explore the space between living rooms and corporations, professional and amateur, reality and utopia." Examples include: Playgrounds Stratalab (an on- and offline-tool for "digital storytelling" that "explores and enables innovative usage, curation and visualization of individual data sets"); Dada Technology (a system of open source hard- and software aimed at making data-sharing more personal and human); "do-it-yourself" production and rapid prototyping (e.g., Palomar5 established the first "fablab" in Berlin in collaboration with "betahaus" to experiment with ways in which everyone can manufacture, repair, create and build products in their own home, thus having control over their design); and Mobile Aid Monitor (a mobile app that makes monitoring and evaluating social funds and activities more transparent and up-to-date).

Multi-Stakeholder Initiatives and Communities

Multi-Stakeholder Initiatives

In recent years, a number of innovative programs have been launched that are explicitly aimed at building partnerships to pool the resources and perspectives of stakeholders in all traditional sectors so as to respond to the enormous complexity of problems around the world. As Kahane (2010) describes it, these programs' leaders and participants each face unusual challenges, including the need to create system awareness, to balance power differentials, and to appreciate and manage differences across cultures and classes. Three examples of this can be found below:

- *Generon Consulting* used its "Change Lab" methodology to initiate the *Sustainable Food Lab*, which aspires to make the mainstream global food system sustainable. The Food Lab is now in its sixth year, with increasing participation and a growing number of projects.[101]
- Together with *Generon Consulting*, the *Synergos Institute* (which later joined Generon in supporting the Food Lab) created the *Bhavishya Alliance*, which attempted to reduce child malnutrition in India.[102] Evaluations of these initiatives document the formidable challenges of taking on such diffuse and complex problems. However, they also point to achievements that validate the potential of partnerships of this kind to open up communication across traditional silos, and to develop collaborative solutions that are worked out by representatives drawn from across the entire system who are all "in the room."
- More recently, *Synergos, Generon, McKinsey & Company* and the *Presencing Institute* formed an alliance with the support of the *Bill & Melinda Gates Foundation* to address problems related to health-care delivery in Zimbabwe.

Multi-Stakeholder Communities

Gerenscer et al. devote an entire book to describing what they call "megacommunities," which they define as "communities of organizations whose leaders have deliberately come together across national, organizational and sectoral boundaries to reach the goals they cannot achieve alone" (Gerenscer et al. 2009: 28). The authors are all consultants at Booz Allen Hamilton, and the communities they describe were all initiated by business-sector organizations. However, megacommunities are by definition conceived of as tri-sectoral, which takes them beyond more limited single-sector approaches such as corporate social responsibility. And they also go beyond public-private partnerships—which are only bi-sectoral—by forging bridges to civil society (ibid.: 62–63). Finally, they are different from multi-stakeholder initiatives because they intentionally form long-lasting ecosystems.

The structure of megacommunities "exhibits many properties of a network," a feature which builds in the capacity for responsibility and adaptability. Gerenscer and his colleagues stress that this is important because "unlike the elegant, self-managed networks of nature, mankind has built an immense amount of complexity into the networks we use every day, and we have not managed them in an effective way. A new degree of connectedness between the diverse component parts is essential" (ibid.: 229). Megacommunities provide that connection in ways that would be impossible without the Web to support.

Just as participants in multi-stakeholder initiatives have found, participation in megacommunities leads to "win-win-win" solutions that "often require transforming the established practices and institutional barriers within their own organizations. For instance, they may need to institute new incentive structures, moving away from those that support individual advancement and toward valuing collaboration" (ibid.: 194). At Enel SpA, the Italian company that the book profiles as a "success story," "social issues have become a strategic item on (the company's) general agenda" (ibid.: 227). The book outlines 10 elements of megacommunity leadership, one of which is long-term thinking. It recognizes that "leaders in some industries have not yet been touched by megacommunity phenomena. But, in our view, we will all be touched by it eventually" (ibid.: 216–17).

Since the Web makes increasingly obvious the value of such ecosystems of communication and collaboration for all sectors, we concur with this prediction. The business sector gains potential business and profits; the social sector gains the opportunity to influence business and partner with it to meet community and societal needs; and the government sector gains the capacity to realize many of its goals without relying on the heavy stick of regulation. Enel discovered the need for megacommunity membership when one of its projects was shut down due to opposition by an environmental group. The story "clearly illustrates the benefits to a specific player within a megacommunity, and the ample benefits that flow to those who work out ways to balance tensions between sectors" (ibid.: 221–222) Indeed, the Web will increasingly provide the infrastructure for the interaction of such megacommunities.

Conscious Capitalism

In our research for this report, we were struck by the number and strength of societal forces contributing to or calling for the emergence of the Commons. To realize its full potential, such a "sector" will likely require special encouragement through government stimulus and support, as Barnes advocates (Barnes 2006: 152–153). However, there is significant momentum in one final arena that we see as likely to add irreversible momentum to the emergence of the Commons. This movement has not come to depend upon or been significantly influenced by the Web. One name for it is "conscious capitalism." But we predict that it will soon bear the fingerprints of the Web as fully as social entrepreneurship does. Indeed, although the latter began long before the Internet became ubiquitous, thanks to the Web, it is now spreading in a seemingly exponential fashion.

The name of Whole Foods CEO John Mackey became famous when it was discovered that he had been criticizing competitors under a masked online identity. Although the case illustrates one of the pitfalls of the Web, Mackey will more likely be remembered for articulating a philosophy of "conscious capitalism."[103] In an exchange with economist Milton Friedman, Mackey wrote (Mackey, Friedman and Rodgers 2005):

> In 1970, Milton Friedman wrote that "there is one and only one social responsibility of business—to use its resources and engage in activities designed to increase its profits so long as it stays within the rules of the game, which is to say, engages in open and free competition without deception or fraud." That's the orthodox view among free market economists: that the only social responsibility a law-abiding business has is to maximize profits for the shareholders. I strongly disagree. I'm a businessman and a free market libertarian, but I believe that the enlightened corporation should try to create value for *all* of its constituencies.

For Mackey, profit is the means, but not the end. This philosophy echoes one that has been practiced for many years by Muhammad Yunus, who won the 2006 Nobel Peace Prize for employing microlending as a strategy for alleviating poverty (cf., e.g., Yunus 2003). It is also similar to the "natural capitalism" philosophy that has long been articulated by Paul Hawken and two colleagues (Hawken 2007; Hawken, Lovins and Lovins 2000). Remarkably, they have been joined by an arch-capitalist, Microsoft Chairman Bill Gates, who used his platform at the 2008 World Economic Forum to make the case for a "new approach to capitalism in the 21st century" (Gates 2008). Among the advocates for a new capitalism is, of course, Peter Barnes (cf., e.g., Barnes 2006: chap. 6, 9), to whom the idea of "reinventing the commons" occurred years before we thought of it, even though our paths to it were independent. Again, we are convinced that the tools of the evolving Web will accelerate the already strong momentum of this idea. Likewise, we believe that this will also give even more impetus to the emergence of a fourth sector, the Commons.

4 Exploring the Leadership Implications for Organization

How should your organization go about positioning itself in this new world? What kind of strategy and approach to leadership is appropriate for it? The temptation is to jump directly to considering which tools make sense to adopt and try out. Granted, there is no denying that there is some value to "jumping in" and gaining experiential understanding of the Web's emerging potential. But organizations would still do well to defer choices about technology until they get a clearer view of the more fundamental questions, such as: What are the most important goals of the organization? How well would they be served by embracing to some degree the cultural shift of which Web 2.0 is just one of the more visible indicators?

Under today's circumstances, there is no single "right" strategy. Indeed, what is right for your organization depends on answers to the question of how it wants to respond to the newly emerging world.

Determining How Your Organization Should Position Itself

Emerging cultural values require a new degree of openness with regard to both information and decision-making. The call for this openness is in most cases too strong to ignore without taking an unwise risk (Weinberger 2008). Societal trends reinforced by the Web are providing pressure for more and more transparency. Thus, for most organizations, and in all sectors, the question is not *whether*, but rather *how* open they need to be to accomplish their overall strategic goals. Those organizations that prefer to be proactive in responding to the emerging realities by defining their own stance, rather than simply reacting, will be well served by striving to embrace new mindsets characterized by interdependence and openness. Charlene Li, who charted the impact of Web 2.0 in a previous book (Li 2008) and has helped a number of organizations adapt to it, stresses the importance of "open leadership" in a recent book of that name (Li 2010). Don Tapscott, author of several books on the Web (including two with Anthony Williams) and the head of a new research initiative charged with exploring the implications of the Web for corporations, reaches the same conclusion (Tapscott and Williams 2006; Tapscott and Williams 2010). As an example of successful implementation of open leadership, Beth Kanter cites the experience of the Indianapolis Art Museum with its public "dashboard" (Kanter 2009e).

Establish Learning as a Foundation

Thoughtful answers to the question of how open your corporation should be require a foundation of learning.[104] In 1992, Peter Senge's blockbuster book *The Fifth Discipline* made the term "learning organization" a buzzword. Despite the popularity of the term and many efforts to apply the concept, few instances have emerged of organizations that have fully developed a learning culture (Darling 2005). The failure has less to do with the relevance and power of the ideas than with the difficulty of changing organizational cultures, which typically reflects the ambivalence of the organization's members toward learning and the tendency of any system to maintain equilibrium. Indeed, there is "immunity to change" at both levels (Kegan and Lahey 2009). Without special training and support, few individuals have the mindsets and skills to productively engage in learning from mistakes without defaulting to finger-pointing. Moreover, few organizations have cultures that enable them to manage their defensive routines in ways that allow members to continuously reflect on how well the organization's behavior is aligned with its goals. Edgar Schein (2002) has taught us that, for learning to take place, the anxiety of not learning must be greater than the anxiety of learning. That perspective makes it less surprising that the best example of a learning organization our research has uncovered is in the U.S. military: the OPFOR units that help prepare troops for deployment in Iraq and Afghanistan.[105]

In the decade since Schein's book appeared, the case for ongoing learning has become even more powerful. During that same period, it has become increasingly obvious that it is more useful to regard organizations as "complex adaptive systems" than as machines (Griffin and Stacey 2005). Such systems must engage in continuous learning about their environments in order to thrive or even survive. Prospecting for new opportunities requires learning, as does striking the right balance between "exploiting" (i.e., sticking with existing products and services) and "exploring" (i.e., searching for new strategies). As Li observes, drawing on her experience with early adapters of Web 2.0, "organizations and their leaders must be constantly open to learning … (and) must learn from employees, customers and partners before they can do anything else" (Li 2010: 168).

Although the relevant mindsets and skills are no easier to learn than they were 20 years ago, social media offer new modes for learning that provide both stimulus and structure for doing so. In our view, the Web makes the possibility of a learning organization more attainable. As illustrated by the examples in Chapter 3, organizations that have begun to adopt social media and other Web-based tools enjoy a strong competitive advantage. Learning about your own organization in terms of employee perceptions can now be greatly facilitated by internal intranets featuring things such as wikis, blogs and community discussion boards. Other Web tools enable "real-time monitoring" of employee and customer perceptions.

Gauge the Desired Level of Openness

Charlene Li makes the case for openness in this way (ibid.: 86–87):

> Social media engagement and financial success appear to work together to perpetuate a healthy business cycle: a customer-oriented mindset stemming from deep social interaction allows a company to identify and meet customer needs in the marketplace, generating superior profits. The financial success of the company in turn allows further investment in engagement to build even better customer knowledge, thereby creating even more profits—and the cycle continues.

Such optimism is grounded in at least some hard data. A 2009 McKinsey & Company survey of 1,700 global executives in a "range of industries and functional areas" asked whether using Web 2.0 was making a difference for their organizations and whether it could be quantified. Sixty-nine percent of executives reported that, by using Web 2.0 technologies, their companies have gained measurable business benefits, including more innovative products and services, more effective marketing, better access to knowledge, lower costs of doing business and higher revenues. Companies that made greater use of the technologies reported even greater benefits.[106]

Still, such benefits are not always evident to those who make decisions. For example, Willms Buhse, a German consultant who is a strong advocate of "Enterprise 2.0" and who has coedited a very insightful set of essays from both sides of the Atlantic (Buhse and Stamer 2008), has conducted many interviews with European managers and consultants. He reports: "There are, of course, always skeptics and doubters." He found this to be particularly true in Germany, a country with a culture not known for taking risks or for its receptiveness to the movement toward open, participatory cultures. CoreMedia, one of the companies profiled in his book (and in our study; see Chapter 3), commissioned the 2007 survey of German companies cited earlier.[107] An analysis of the findings by Nicole Dufft reported that (ibid.: 144):

> (The) usage of Web 2.0 applications within companies is very sobering: such tools are primarily used—if indeed at all—only by individual members of staff. Only a fraction of the companies surveyed have so far implemented Web 2.0 applications company-wide or even across departmental boundaries. ... (While)

70 percent of the companies have an intranet system that is provided as a company-wide service ... only 10 percent of all companies have established these as cross-department or company-wide services. Even internal blogs and wikis for staff and projects have so far rarely been implemented by the surveyed companies as services that venture outside department borders.

The author goes on to comment: "Given this environment, one finds, unsurprisingly, that two-thirds of respondents cite the unclear business benefit of Web 2.0." In fact, only one-third agreed with the proposition that Web 2.0 applications "will be part of daily company business in a few years," and even fewer (25 percent) thought that Web 2.0 "will make processes and collaboration more efficient." If this were not so-bering enough, she speculates: "Without substantial networking support ... wikis, blogs or social bookmarking tools will serve only to create additional knowledge silos, which then hinder rather than help the efficient use of knowledge and information" (ibid.: 145). This lack of appreciation of the benefits of the Web can only reinforce the barriers posed by national and organizational cultures. As a senior project manager at the Bertelsmann Stiftung observed: "Although many aspects of leadership require an extremely high level of openness ... (it is) always considered as a weakness by Leadership 1.0 institutions."

Our reading of the data suggests that, on average, the benefits of the Web are quite high in relation to costs and that supporting data will only continue to mount. However, each organization will need to make its own determination. They will have to ask themselves what degree of openness would best serve their particular organization. In doing so, it's useful to assess the benefits and weigh them against estimated costs and risks.

Gauging the Benefits of Openness

The examples in Chapter 3 are all drawn from organizations that found value in opening up access to information and decision-making. It may be helpful to look back at that chapter, being on the alert for examples that might serve your organization's objectives better than your existing strategies. People in organizations wedded to ROI will be most comfortable with the examples that have generated tangible results. Li is very optimistic on this point (Li 2010: 86, endnote 10): "(We) found there was correlation between deep, broad engagement and financial performance, specifically in revenue and profit. Companies that are both deeply and widely engaged in social media surpass other companies in terms of revenue, gross margin and gross profit performance by a significant difference."

Recognizing that correlation does not prove cause and effect, Li goes on to calculate estimated ROI for several strategies based on hypothetical examples (e.g., open dialogue, open support through information sharing, open innovation). Her ROI esti-

114

mates range from a low of 150 percent to a high of 1,667 percent. The highest esti-mate reflects a very high estimated impact of open dialogue on the quality of relation-ships with customers (ibid.: 89–100). Although we are not persuaded of the accuracy of the precise numbers that Li generates, we see compelling evidence that she is pointing in the right direction, as her findings on the value of openness are confirmed by others (cf., e.g., Tapscott 2008a; Tapscott and Williams 2010). A culture of greater openness regarding information sharing and decision-making generally pays off. The question then becomes: What are the costs and risks?

Gauging the Costs and Risks of Openness

Many of the perceived risks of openness have to do with letting go of control, which is often seen as *losing* control. While it's important to recognize that this could happen, it is also important to consider how much control one really has in the first place. In environments of rapid flux and high unpredictability—that is, precisely the environ-ments in which Web-supported open leadership strategies are most useful—a sense of control is probably illusory. As Clay Shirky (2008) has written: "The loss of control you fear is already in the past. ... You do not actually control the message, and if you believe you control the message, it merely means you no longer understand what's going on." In fact, as the contributors to a recent collection of articles on implement-ing Enterprise 2.0 attest to from multiple perspectives, the "art of letting go" can ac-tually result in *more* control—or at least more influence—rather than less (Buhse and Stamer 2008).

Of course, there are organizations that do not need to be more open to succeed. An oft-cited example is Apple, which has had remarkable success despite apparently being closed and controlling. Would that work for your organization? Perhaps it would if it has what Li calls the "Apple factor"—"a combination of brilliant engineers and designers, a charismatic CEO and a brand that everybody loves" (Li 2010: 71).[108] In any case, an "openness audit" could be helpful in arriving at your answer (ibid.: 44–48).

Choose an Appropriate Structure to Support the Strategy

Which structure would best support your strategy? Again, there is no right answer. But some structures will likely serve the level of openness you choose better than others. A common approach, which can be successful and which leaves top executives in their comfort zone, is a centralized structure. Companies as diverse as Starbucks and Ford are examples. A variant of this approach that is more open is illustrated by Cisco Systems, which (as our case profile in Section 3 shows) uses a structure of deci-sion-making that is distributed.

An even more open approach, but one that still retains a degree of centralization, is to set ground rules and constraints at the top and then to let solutions bubble up without trying to control them. Snowden (2008) calls this "top-down stimulation of bottom-up activity." Such approaches leverage the power of a new mode of production, peer production, which Tapscott defines as "a way of producing goods and services that relies entirely on self-organizing, egalitarian communities of individuals who come together voluntarily to produce a shared outcome" (Tapscott and Williams 2006: 67).

The most dramatic forms of this are illustrated by two organizations that have evolved out of the open source software culture, Wikipedia and Linux. Indeed, posting articles to Wikipedia and contributing code to the Linux operating system are examples of peer production. In the unlikely venue of *Foreign Affairs*, Anne-Marie Slaughter makes the case that the whole world is becoming networked, observing (Slaughter 2009: 97): "(Under) a system of peer production, supply chains become 'value webs,' in which suppliers become partners and, instead of just supplying products, actually collaborate on their design." She cites Boeing as "a particularly striking example" of peer production, noting how it "has shifted from being simply an airplane manufacturer to being a 'systems integrator,' relying on a horizontal network of partners collaborating in real time. They share both risk and knowledge in order to achieve a higher level of performance. It is not simply a change in form but a change in culture. Hierarchy and control lose out to community, collaboration and self-organization" (ibid.).

The same basic idea can be implemented in more traditional organizations using a "coordinated" approach. It retains strong centralized direction (regarding policies, practices and perhaps preferred technology platforms). But each department or staff chooses how it will achieve goals and launch initiatives (Li 2010: 149). Li has seen this approach work well under several conditions. It "is well suited for decentralized organizations that want to create greater synergies and collaboration between various efforts," and at the same time, one that even "more mature organizations tend to gravitate toward, as they seek to spread best practices throughout the organization" (ibid.) The American Red Cross (profiled above) is another example. And this is also where Hewlett-Packard started in 2005. HP insured that all officially sanctioned blogs had the same "look and feel" and an Internet address that included "hp.com." Content, however, was handled by individual bloggers. Blogs soon sprang up throughout the organization, along with a "center of excellence" at the corporate level. The disadvantage of this approach is that the organization will not move as quickly as it would with centralized approaches. Likewise, the practices disseminated through the blogs may not always be the best.

Another alternative is what Li calls an "organic" approach that allows openness to develop where it is most natural (ibid.: 145). Microsoft is one example of a company following this approach. Anyone in the company can blog or put up a Facebook page. Another instance is Humana, an American Fortune 500 health insurance provider. It uses a "town square" in which each business unit and department sets up its own "social media outpost" (ibid.). Li quotes Greg Matthews, Humana's director of consumer innovation, as saying (ibid.: 146): "There was recognition that every piece of our business knows how to do its business best, and is going to be able to design a social media function that is best able to support that strategy. We didn't want this to have a top-down control structure." Although Li found the results of this "impressive," the risks of such an approach are not trivial. It will not necessarily lead to a concerted push into the new world of the Web, and it can in fact lead to confusion, inconsistency and lost productivity. Nevertheless, Li feels that it is worth the risk, and finds that it is "ideally suited for companies just beginning to venture into social media and openness, who want a flexible approach that taps into the enthusiasts and needs that already exist in the organization" (ibid.).

Answers to the question of how open to be will naturally vary. But what remains constant will be the value of asking the question. Indeed, if your organization is not asking that question, how can you encourage it to do so?

Encouraging Your Organization to Respond Strategically

If you are like most managers, your organization does not resemble the examples you will find in Chapter 4 of organizations on the cutting edge with respect to Web technologies or open styles of leadership. Likewise, you need to ask yourself whether your organization should really be on that edge, since after all, to be a relatively early adopter is to take risks. The best way to decide is through undertaking the rigorous engagement with that question needed to make a conscious choice. It is wise to avoid the risk of encountering a situation in which it is already too late to catch up by the time you know you need to be doing something different.

Assuming this makes some sense to you, what can you do to encourage your organization to be strategic rather than reactive? First, it is reasonable to assume that present organizational practices are in place because they have worked to some degree. More fundamentally, those practices are an expression of less visible, but deeply rooted mindsets and assumptions that constitute an organizational culture. As we have already observed, organizational cultures are highly resistant to change, and in the long run, they will only change as the organization demonstrates that other ways of "doing things" work (Schein 2010). In the short run, there are seven steps worth considering:[109]

1. Gain personal "Web literacy" and foster its acquisition by your team.
2. Encourage a long-term thinking process that addresses Web strategies.
3. Encourage your organization to develop policies on the use of social media.
4. Encourage someone in the C-suite of your organization to use social media.
5. Help your organization anticipate/address common barriers to open leadership and the adoption of Web tools.

6. Encourage your human relations, marketing and communications departments to experiment with social media.
7. Ensure development of Web strategies from multiple perspectives.

In what follows, we will consider each of these in turn.

Gain Personal Web Literacy and Foster Its Acquisition by Your Team

When we consulted with Web experts in the course of our study, the most consistent recommendation they made was to encourage managers to "jump in" and gain first-hand experience with social media. In fact, we found that we had to take that advice ourselves.[110] Merely reading about it or reviewing lists of best practices—however inspirational they may be—is no substitute for hands-on experience. Trying it out in your personal life is a sensible and very low-risk first step if you have not already done so. A next step would be to explore professional communities of practice, which will give you a taste of how the Web is enabling new forms of peernetworking and learning. These experiences—if they don't discourage you—will enable you to speak with confidence should you decide to encourage others in your organization to dip their toe in the water. (The bonus is that such exploration will provide an opportunity for relationship-building with any children in the family, as you invert the usual coaching hierarchy!)

In the unlikely event that no one else on your team is more experienced than you are, such exploration will also enable you to serve as a role model. In any case, you will be a role model for learning, and encouraging your immediate team (and peers) to gain experience is a good practice field before trying out the next steps.

Meister and Willyerd confirm this advice and conclude that "you must be a user of social media to transform your business." To do so, they suggest taking the following steps (Meister and Willyerd 2010: 151):

- Joining and participating in several major social networks, such as Facebook, Twitter, LinkedIn and YouTube;
- Conducting a search to find out how your competitors are using these social networks and reporting back to your team on your findings;
- Learning the language of social media by reading glossaries of books on the subject.

Beth Kanter also recommends finding a colleague ("study buddy") to learn with.[111] Indeed, the principal author of this study got great value from taking several Web-based courses in social media.[112]

Encourage a Long-Term Thinking Process that Addresses Web Strategies

For an organization to make an informed choice about the best stance to take regarding the Web and its culture, a planning process that initially—or simultaneously—establishes the organization's most important priorities is almost absolutely essential. At the same time, it's worth noting that the acceleration of change in recent years raises serious doubts about the value of approaches to strategic planning that stress having a vision supported by long-term goals. Indeed, as early as 1994, Henry Mintzberg wrote a book supporting this view entitled *The Rise and Fall of Strategic Planning*. Not surprisingly, 15 years later, things are changing too fast for any prognostication to be more than just a good guess. One of the experts we consulted, the project manager of the "Future Challenges" initiative at the Bertelsmann Stiftung, affirmed the wisdom of skepticism regarding such planning. Nevertheless, he still concluded that it was of critical importance "to have an occasion for in-depth dialogue on the risks and benefits of using social media in the context of a generative conversation about what the organization's most important priorities are and how best to achieve them."[113]

Still, the growing skepticism about *planning* should not discourage efforts to base organizational decisions on long-term *thinking*. Indeed, scenario planning has demonstrated a value that is only likely to increase along with external complexity and uncertainty (Schwartz 1996).

A very recent study by Meister and Willyerd offers a helpful map. It was based on a survey of 2,200 working professionals from around the world about their expectations of employers, a survey of 300 such professionals on their current and expected practices, and more than 50 case studies. In the end, the authors recommend the following seven steps (Meister and Willyerd 2010: 141–146):

1. *Identify business drivers* (Improved decision-making? Increasing employee engagement? Reducing time to market? Attracting "Millennials"?).
2. *Form a coalition of stakeholders* (e.g., a "cross-functional collaboration encompassing the corporate learning, HR, IT, legal and internal communications departments").
3. *Host a social media boot camp* (e.g., create a learning experience in which the team actually uses various types of social media to share ideas).
4. *Create a launch plan* (e.g., by clarifying three critical roles: those of community moderators, community administrators, and internal marketing and communications experts).
5. *Develop a pilot offering* (perhaps using one department as a pioneer,[114] or one business challenge[115]).
6. *Design a communication plan* (to dispel common myths and concerns[116]).
7. *Agree on metrics* (both qualitative and quantitative metrics in categories such as internal processes, customer-related processes, and external partners and suppliers).

Above all, they emphasize "the importance of viewing this as a business initiative rather than just an HR or learning one."

Encourage Your Organization to Develop Policies on the Use of Social Media

An area in which it is important to be strategic rather than reactive is that involving the clarification of policies regarding use of social media for personal and business purposes (cf., e.g., Kanter 2010d). This discussion can serve as an indirect stimulus for step three if there is not a more immediate path. Such policies provide structure around the degree of desired openness. Simply "putting our faith in people and letting them do what's right" can result in an inconsistent and ultimately incoherent set of policies and practices (Li 2010: 106). And just like any new relationships, those created through the Web require that etiquette be observed. Indeed, experienced organizations communicate expectations regarding the use of the new tools for openness and bolster them with explicit policies and procedures.

Examples can be helpful. Li gives several, along with a "Social Media Guidelines Checklist" (ibid.: 112), and links to the full text of the guidelines from a broad range of organizations. Meister and Willyerd provide a detailed summary of Intel's "social media rules of engagement" (Meister and Willyerd 2010: 147–149). There are many other sources for such "acceptable usage policies." At the unstructured extreme end of the continuum is Zappos, the online shoe and apparel company, which has no policies at all, but instead relies on a rigorous training program to instill company values. Similarly, Microsoft has allowed employees to begin blogging, constraining them only with an informal policy consisting of calls to respect confidentiality and be "a rational, thinking person."

By contrast, most organizations are far more distrusting. A 2009 survey indicated that slightly over half of companies block use of social media sites altogether.[117] Likewise, some organizations use a "managed service" to control employees' access to the Web, for example by allowing access only between certain hours. Li feels that significantly limiting social media use by employees is "the wrong approach," and we concur, believing that in most situations it makes sense to err on the liberal side by having relatively permissive social media policies. Indeed, as the example of Best Buy illustrates, initial use of social media in the workplace has sometimes triggered the emergence of practices that make significant contributions to company goals and profitability. The American Red Cross found it helpful to go beyond simply having a policy, developing a "social media strategy handbook" (Kanter 2009d).

In this context, it is useful to remember how Hagel and his coauthors stress the ways in which knowledge has changed in recent years, from placing an emphasis on "stocks" (e.g., books, libraries) to one on "flows" (bits of knowledge conveyed in conversations or accessed "just-in-time"). They write (Hagel III, Brown and Davison 1999: 56): "(If) firms want to enhance their participation in tacit knowledge flows, they must find ways to expand and enrich the social networks of their employees, helping them to connect with other individuals on relevant edges." Social media exponentially increase the power of social networks, and the access of employees to social media from and in the workplace can magnify the benefits even more.

Of course, as Willms Buhse reminds us (Buhse and Stamer 2008: 139): "The risks should ... not be underestimated: Via blogs, staff members will begin to inform the public about technical, organizational and even personal details—content that may even be sensitive information about the company." The already discussed cases of Eurostar, British Airways, and Virgin Atlantic provide vivid examples of these risks. At the same time, managers who are concerned about loss of productivity would do well to keep in mind Beth Kanter's finding that: "Millennials say that if you cut them off from Facebook or LinkedIn during work hours, it is like cutting off an arm. It isn't a waste of time—there is such a thing as social productivity."[118]

Encourage Someone in the C-Suite of Your Organization to Use Social Media

An early step that has worked in many situations is having the CEO put a visible foot in the water by launching a blog. Earlier, we cited the example of Paul Levy from the Beth Israel Deaconess Medical Center. Styles can vary widely, ranging from the deeply personal (e.g., that of Matt Blumberg, the founder of Return Path) to the highly opinionated (e.g., that of Mark Cuban, the owner of the Dallas Mavericks) to the purely business-oriented (e.g., that of Bob Lutz, the former vice chairman of General Motors). But the payoffs can be quite tangible. Bill Marriott, the CEO of Marriott International, attributes $4 million in additional business to his blog. At least 60 CEOs are currently blogging, along with 300 other senior executives.[119] In fact, one forecast suggests that, by 2020, "job requirements for CEOs will include blogging" (Meister and Willyerd 2010: 220).

These data suggest that a blog coming from the C-suite can effectively serve multiple objectives, including those of:
- Overcoming a sense of remoteness from the top;
- Increasing transparency within the organization;
- Encouraging openness by inviting comments and feedback; and
- Role-modeling the effective use of social media.

Blogging is, of course, only one way to engage with social media. Zappos CEO Tony Hsieh is a fan of Twitter, where he had nearly 180,000 followers in January 2010. In fact, he credits Twitter with helping him grow personally (ibid., quoting a blog post of Hsieh from Jan. 25, 2009):

> By embracing transparency and tweeting regularly, Twitter became my equivalent of being always on camera. Because I knew that I was going to be tweeting regularly about whatever I was doing or thinking, I was more conscious of and made more of an effort to live up to our 10 core values.

Help Your Organization Anticipate/Address Common Barriers to Open Leadership and the Adoption of Web Tools

The most important likely barrier is also the hardest to address: organizational culture. As we have seen, Web 2.0 represents a shift even more difficult—and far less tangible—than replacing an organization's legacy IT infrastructure. Despite the prevalence of tools that purport to help organizations assess their culture, it is actually very hard for members of an organization to identify their own culture, even with the help of tools designed by well-intentioned experts. Ed Schein, one of the foremost authorities on this subject, believes that such tools primarily assess organizational "climate," a much more superficial phenomenon.[120] In any case, it would be difficult to assess the culture as a whole, which contains numerous assumptions about matters as fundamental as space and time (Schein 2010). Schein spent two decades as a consultant to the leadership of Digital Equipment Corporation before he felt comfortable making pronouncements about the firm's culture (Schein 2003).

Is there anything you can do to address potential cultural barriers? Absolutely. Without great investment of time or money, you can identify the features of the culture most likely to affect the aspiration in question and state it in terms as concrete as possible. Schein describes the design for a half-day meeting involving people experienced in the organization to identify cultural "artifacts" (policies and rituals), espoused values, and underlying assumptions and values (Schein 2010). With this information, one can prioritize building upon and reinforcing supportive assumptions and practices, while addressing some of those most likely to get seriously in the way. Beth Kanter (2010i) also offers a complementary alternative approach consisting of seven steps for confronting social media fears. Although it is designed for nonprofits, it is applicable to any organization.

Ultimately, the biggest barrier to culture change may be the individual mindsets held by organizational members as a function of their development stage or "action logic" (Joiner and Josephs 2006; Torbert et al. 2004). Only a small percentage of people in leadership positions are likely to have developed to the point where they bring assumptions to their role that are consistent with values such as openness, transparency and collaboration. As we have seen, most of us instead have a default tendency toward more "unilateral," self-protective governing values that drive our behavior—our "theory in use"—regardless of our espoused values (Argyris, Putnam and Smith 1985). Andrew McAfee acknowledges that the deepest long-term threat to Enterprise 2.0 "comes not from managers who don't want more truth and franker discussion within their organizations, but rather from those that sincerely do" (McAfee 2009a: 200). At the same time, McGuire and Rhodes (2009) have shown that aspirations to create a culture in which interdependence and collaboration are the norm can provide a supportive context that is itself a lever for transformational change at the individual level. Matthew Mezey (2010) provides a very current and sophisticated review of both the potential of—and the barriers to—bringing open leadership to Enterprise 2.0.

The deepest barriers to new forms of leadership and the exploration of social media are cultural. But in the form of tacit assumptions about "the way things are done around here," there are often specific organizational routines that get in the way as well. For example, there may be legal constraints that are a barrier (as, e.g., the Obama administration discovered when trying to encourage use of social media by federal departments). Likewise, organizational equivalents can seem like caricatures. For example, one expert told us in confidence that his or her organization's legal department insisted on a formal contract with anyone writing a blog. Why? To ensure that the company owned the rights to the information! Such examples are perhaps not surprising in light of the distrust that has prompted more than half of all companies to ban use of personal media at work. But if such policies and practices are not rooted out and changed, they can easily kill fledgling efforts to bring social media into the organization.

In our experience, other common barriers are easier to address:

- Inflated expectations about the speed or degree of payoff from social media tools.
- Lack of internal understanding of what it takes to work collaboratively.[121]
- Lack of (or underestimation of the demand for) user support, which can lead to early frustration with attempts to use the tools (e.g., log-in problems).
- Overly complex tools with too many features and options.

For high-profile initiatives, such as rolling out an intranet or other internal platform, it is worth investing in substantial prior thought, even to the point of identifying opinion leaders and getting them to commit to participating. The experience of the Open Society Institute (OSI)—headquartered in New York City, but with satellite offices and programs scattered throughout the world—is illustrative of a number of these points. OSI engaged experts on organizational change to help plan the launch of KARL, an internal platform for collaboration designed to break down barriers among different programs and departments. Program leaders attribute the ultimate success of the program to careful planning of this kind as well as to the fact that the technology was kept very simple.[122] This has led an increasing number of organizations to adapt OSI's platform instead of more complex (and expensive) options, such as Microsoft SharePoint. Indeed, this is one of the many reasons for being cautious about letting IT people take the lead. So-called tech geeks typically love advanced features and lots of options, leading to tools that are not friendly to the average, less tech-savvy user.

Encourage Your Human Relations, Marketing and Communications Departments to Experiment with Social Media

A time-honored approach to change management is to begin with "low-hanging fruit" or "easy wins." These are often to be found in HR, marketing and communications departments. In fact, they may even be ahead of you, as such departments are often the parts of an organization most likely to see a tangible personal benefit from using Web 2.0. In Chapter 3, we cited examples of companies that have used the Web to take advantage of its radically lower advertising costs. In her survey of German companies, Dufft found that professional staff in the externally oriented marketing and communications departments often take the lead in recognizing and exploring the benefits of social media (Buhse and Stamer 2008: 148).

HR departments in particular can play a powerful role in demonstrating the power of Web 2.0 internally. For example, they can use social media to create ways for employees to discover common interests (including prior experience) and share information (e.g., by posting reports on a conference attended). More generally, HR personnel will do well to look for opportunities where they can show a quick payoff to managers by introducing Web tools and thereby provide a positive entry experience. Many HR departments have already turned to the tools of social media for recruiting and been rewarded for doing so by the results achieved. Given access by such bridgeheads, new tools can slowly but surely encroach on internal company processes until they eventually become a normal part of the business environment.

Practices and tools of benefit from an HR perspective are well documented. *Human Resource Executive Online*, the blog that summarized the results of the 2009 McKinsey & Company survey mentioned earlier, also cited the experience of a vice president at EMC, a global data-storage and security-solutions firm based in Hopkinton, Massachusetts, who said: "The people that are using Web 2.0 are getting serious benefits. … It's really inevitable that these technologies will have an impact." Just a few examples of the employee benefits cited are that:

- Web 2.0 helps the workforce meet the human need for connection, respect and a sense of purpose.
- Sparked by the 2.0 behavior model and tool set, employees are doing great things for their personal and professional development.
- Workers are engaged in collaborative communities formed around interests (e.g., art, green issues, running, innovation, culture, etc.).
- Network users are sharing, among other things, favorite business books, restaurants and company benefits.
- Employees are also being heard more—and more quickly—by senior management. A "water cooler" community takes employee input and comments and puts them in a place where everyone can weigh in and respond.

"The employee suggestion box is a dinosaur," the vice president added, noting that social media are "also inexpensive … as you can get a lot of the applications for free." In the end, she concluded that Web 2.0 "is helping build mutual trust, understanding and engagement" (Starner 2009).

A recent book adds further confirmation—and many examples—of the value of Web 2.0 for the workplace. Of "10 forces shaping the future workplace now" identified by Meister and Willyerd, two involved Web technology ("digital workplace" and "ubiquity of mobile technology"), three were a result of such technology ("culture of connectivity," "social learning," "the participation society") and another featured the new generation that takes such technology for granted ("Millennials in the workplace") (Meister and Willyerd 2010: 13–40).

In particular, the authors stress the power of six "meaning motivators" for this generation, most of which are enhanced by Web 2.0 (ibid.: 93–119). They found that Millennials want to:

- Make a difference in the world;
- Feel they are contributing in the workplace;
- Be innovators;
- Be heard;
- Know they are succeeding; and
- Express who they are through work.

Looking ahead, some observers predict that, within a decade, "the lines among marketing, communications and learning will blur." They advise organizations to "look for partnerships among heads of human resources, corporate learning officers and chief marketing officers as corporate training programs are re-imagined as consumer education online offerings and become part of the marketing/communications mix to increase market share and consumer satisfaction" (ibid.: 229).

Expecting departments such as HR, marketing and communications to be pioneers in using Web 2.0 may seem counterintuitive. In what follows, we discuss whether it is reasonable to expect IT departments to play this role.

Ensure Development of Web Strategies from Multiple Perspectives

Experience to date suggests that organizations relying only on their IT departments to identify new technology will lag behind those who empower all employees to explore for themselves. This is important not only in small organizations that may not have the resources to employ dedicated and highly expert IT staff or that have IT departments that are struggling just to keep a network alive. From what we have read and heard from Web 2.0 pioneers, even well-staffed IT departments are unlikely to take a leadership role. Indeed, though the situation may vary substantially, there is reason to fear that such departments may take a suboptimal or even defensive approach to adapting to the Web. IT-driven solutions may be governed not only by turf considerations, but also by deeply ingrained professional habits of mind—which might be more understandable, but are also still harder to counteract.

Even when an IT department wishes to be supportive, it may not know the best way to do so. In one story we heard from a person wanting to set up a wiki to enable virtual input on a document, although IT staff were quite willing to help, they still took weeks to come up with a solution, and the configuration they ultimately designed made it hard for users to make comments and hold a discussion. The result was a frustrating and time-wasting experience. The storyteller told us that he then consulted Web-savvy experts on how they would solve this problem. This resulted in a quick, easy, user-friendly and free solution. For this particular purpose, they recommended a simple combination of posting a document on Google Docs and holding a conference call. Our informant concluded that there is a world of difference between the experience and mindsets of "IT people" and "Internet people."[123]

Based on his own experience, Bertelsmann Senior Project Manager Ole Wintermann commented on this distinction: "Web people are interested in content, in moving things forward. IT people are more focused on tools. Web 2.0 is more about mindsets, but the IT department thinks IT is the essential part. Web 2.0 people see things emerging from chaos. IT people are structured engineers, working from a manual. Web people challenge the limits that the IT people see. But what is important is that both groups are needed for getting a better result through cooperation."[124]

In time, of course, IT departments will evolve in ways that are synchronous with the requirements and opportunities of the Web. In fact, some already have. Mulholland, Thomas and Kurchina describe a "major shift" currently underway: the evolution of "edge" IT, which is entirely different from traditional "hub" IT. In the hub approach, traditional IT applications—such as ERP (enterprise resource planning) software—are seen as residing at the center and as being connected to the front lines by spokes consisting of stable business processes. On the other hand, "edge" or Web 2.0 IT represents the organic efforts by IT staff to develop tools—frequently ad hoc— to support employees (Mulholland, Thomas and Kurchina 2007: 14).

5 Cautions About the Leadership Impact of the Web

Needless to say—but nonetheless important to acknowledge—is the fact that the enormous potential of the Web does not come without costs and risks. Indeed, as Clay Shirky notes (2008, cited in Thompson 2006: 11): "The normal case for social software is failure." Likewise, success can come at a high price. According to the experience of Kasper and Scearce, "as with using any tool for the first time, social media experiments often require a steep learning curve that can be quite time-consuming" (Kasper and Scearce 2008: 9).

The chief obstacles are not technical. Some have to do with understanding and expectations. A *BusinessWeek* article identifies and debunks a number of misleading "myths" about social media (e.g., that "social media is cheap, if not free" and that "you can make a big splash in a short time") (Ochman 2009). But the obstacles are more than informational. As one consultant who has worked with the U.S. Marine Corps to help it adopt Web 2.0 has written (Singh 2009): "We can install a Web 2.0 system in an agency probably within a few hours or a few days. But the move to Web 2.0 is about behavior change, organizational change and changing the mindset of what collaboration is."

There are risks as well as costs. No one can deny that newspapers as we know them are in a state of demise and that radical changes are underway in book publishing. Whether these developments are to be welcomed or regretted is a matter of opinion, but some real and perhaps irreplaceable losses seem evident (e.g., in-depth news reporting and investigative journalism). Among the other concerns most frequently mentioned are:

- Information leakage (whether it is proprietary, confidential or simply not meant to be shared) and the erosion of information quality (e.g., because Web data is not always vetted);
- Information overload (i.e., having too much information for it to be effectively processed) and "tool fatigue" (i.e., having too many tools to keep up with);[125]
- Loss of business productivity through personal use of social tools during the workday;

- Erosion of personal space and time and the quality of consciousness;[126]
- Threats to privacy from social media and new tools for "enhancing" user options;[127]
- Threats to individual intelligence through multitasking (Carr 2010) and to collective intelligence through the emergence of "hive mind" (Lanier 2006);
- Threats to authority and genuine expertise when everyone has a voice;[128]
- "Creative destruction" (e.g., unemployment and the demise of organizations);[129]
- A "digital divide" that exacerbates social inequality; and
- The Web's potential used for ill as well as good.[130]

The weighting of these factors will of course vary by national and organizational culture as well as by individual preferences and comfort zones. We were often told that since managers in Germany, for example, are quite conservative regarding protection of data and intellectual property, they are also cautious about allowing access to social media during working hours or in making the shift to the cloud. Silicon Valley firms in the United States, by contrast, tend to be liberal in their policies on the use of social media. Thus, it is really up to each organization—and individual—to make a judgment about how to fully embrace the tools of the Web in light of these risks and costs.

At the same time, we reiterate that it is critical to recognize that there is increasingly little choice about whether to deal with an emerging new reality. In this reality, generational norms that have been reinforced by the Web are inexorably shifting in the direction of youth—toward more openness and participation as well as toward less hierarchy. Under these circumstances, some degree of risk and cost appears inevitable. But the question remains as to whether it is worthwhile. On this question, there is no lack of either optimists or pessimists.

In the next and last chapter, we summarize key points and make clear our own—and admittedly subjective—view.

6 Concluding Reflection

The evolving Web is the source of new technologies that are perhaps the most tangible of many changes that are transforming society—and organizations within all sectors—in unforeseen and unprecedented ways. Indeed, the world is becoming more complex, more interdependent and less predictable. Still, these new technologies are only the surface manifestation of a deeper shift, one that has to do more with culture, as received norms are challenged to absorb practices of transparency, collaboration and openness. These practices, which emerged in the "geek" worlds of open software, have become second nature to the Millennials who grew up digital. This shift will challenge organizations, and those who exercise leadership within them, to understand the new conditions and make accommodations that are appropriate to the context. At the same time, new modes of leading and new tools for the exercise of leadership come hand-in-hand with the new constraints on leadership and culture. The same technologies that threaten to make traditional ways of leading obsolete offer powerful new vehicles for innovation and change, eroding boundaries around and between organizations while simultaneously fostering ecosystems that mitigate risk and facilitate creative adaptation.

In the new world, as in the old one, there is no single right way to exercise leadership. Instead, what is right depends on the context. But in choosing the best approach, those wishing to exercise leadership will experience strong pressures in most settings to be more inclusive and participative, and to learn to build and leverage networks even as they continue to rely on formal roles and hierarchy. The temptation will be strong to deny and resist these pressures rather than to accept and embrace them. However, those who are able to take the risk of "letting go" may discover that they gain enormous opportunities to learn and extend their influence—while losing little more than the mere illusion of being in control. They may realize the possibility of tapping more fully into the potential of those around them, and of mobilizing greater commitment and alignment in support of the directions that emerge.

These are not easy times to be in formal positions of management and leadership. Guardians of organizations at all levels face tough choices about how much to insulate and protect their institutions from the threats to privacy and security posed by the Web while at the same time striving to benefit from the Web's power to open access to new ideas and modes of organizing. More fundamentally, organizations of all kinds face challenges to their viability as they strive to keep pace with the agility and cost advantages of Web-enabled networks and free agents. Indeed, creative disruption may become the new status quo, and whether one focuses more on the disruption or the creativity may depend as much on personal disposition as on one's particular location in an organization, country or culture.

We end as we began, with a highly subjective reflection: Thanks to the Web, we have the opportunity to learn how to hone and extend our individual intelligence, deepen our collective intelligence and use this new capacity to address threats to our well-being and survival that have resulted from the accumulated, unintended systemic consequences of our behavior. Thus, the ultimate implication of the Web for leadership is that it provides hope for a sustainable future combined with the tools to help create it.

Appendix: Trends in the Evolving Web[131]

Since the emergence of the World Wide Web in the early 1990s,[132] an array of technologies and tools has evolved at an exponentially increasing pace. These tools have radically expanded the possibilities for communication and interaction at all levels of society—and they promise extraordinary change. As Shirky puts it (Shirky 2008: 107): "When new technology appears, previously impossible things start occurring. If enough of those impossible things are important and happen in a bundle, quickly, the change becomes a revolution." Indeed, the creation of the Web[133] has been compared to the invention of the printing press. As Shirky describes it (ibid.: 20–21): "We are living in the middle of a remarkable increase in our ability to share, to cooperate with one another and to take collective action, all outside the framework of traditional institutions and organizations."

In this section, we will explore the basic features of the Web as a first step toward understanding the implications for leadership held by these revolutionary new technologies. Although the tools that have emerged via the Internet are fostering momentous change in society and organizations, they are just the latest phase of an old pattern whereby technological change both triggers and drives social change. Below, we provide a helpful overview of the evolution of this technology and identify the specific tools that were available at specific stages.

Pre-Internet Tools

For decades, we have relied on technology to enhance the effectiveness of organizations and leadership. The most common of these are tools so familiar that they need no definition:

- Telephone (and telephone conference calls)
- Television
- Audio- and videotapes

- Videodisks
- Computer-assisted instruction and simulation games
- LCD projectors

These tools had powerful implications for leadership and learning. The relatively limited impact of these tools only became apparent with the paradigm-shifting appearance of the Web.

Web-Based Tools

Most observers of the Web find it useful to divide its evolution into various phases. However, they tend to disagree about whether to do so simply on the basis of chronology, or rather on the appearance of distinctive features. Since we see virtues in both approaches, we will draw on each of them.

One active blogger on the topic offers a timeline suggesting 10-year units of demarcation:

Figure 2: Timeline (2008)

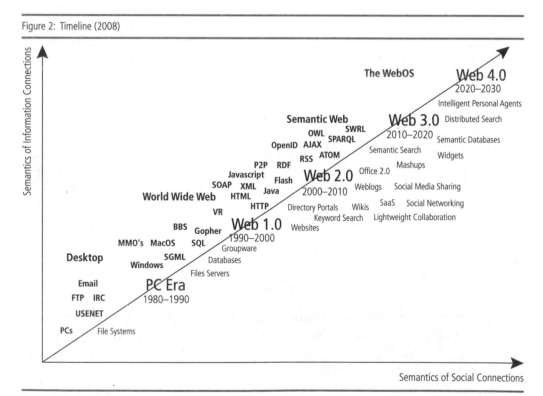

Source: Radar Networks & Nova Spivack, 2007—www.radarnetworks.com

Many of the tools featured in this timeline will be unfamiliar to anyone not deeply immersed in these technologies, and not all these tools are cited in the report. However, the timeline reflects the conviction of the author, Nova Spivack, that the complexity of Web activity makes it difficult to define phases by how they differ from one another. Thus, he argues that it is better to simply use a timeline in which Web 1.0 and Web 2.0 each last a decade. We agree that the two phases that have emerged so far have lasted roughly a decade each, but we also see value in describing how those phases differ from one another and in anticipating what is to come.

Web 1.0

In its first decade, the Web significantly broadened the ease and range of communications. Web 1.0 is characterized by a tendency toward the one-way broadcasting of information, though some participatory media did emerge. Some of these tools—such as e-mail and instant messaging—are also now so familiar as to require no explanation. Other tools that are familiar to many, but probably not to all, include the following:

- *Chat rooms* are any type of synchronous (or occasionally even asynchronous) conversations conducted on the Internet.
- *E-learning* is a type of learning using technologies that are mainly Internet- or computer-based to reach learners. In some instances, no face-to-face interaction takes place. It is typically implemented through a so-called learning management system that employs software for delivering, tracking and managing training.
- *Listservs* are electronic mailing-list management tools. The name is based on the first such tool, LISTSERV.
- *Podcasts* are any combination of software and hardware that permits the downloading of audio files (most commonly in MP3 format) for listening at the user's convenience. The term was inspired by Apple's iPod. Professional broadcasters and syndicated radio shows are starting to make their content available as podcasts.
- *Search functions* are various means of conducting searches via the Internet, such as the key word searches employed by Google.
- *Simulation games* are reenactments of various activities of "real life" in the form of a game. Simulation games serve a wide range of purposes, including training, analysis and prediction. Well-known examples are war games, business games and role-playing simulations. They may be enhanced by computers.
- A *text message* is a short (160 characters or fewer) message sent from a mobile phone using the Short Message Service (SMS). It is available on most digital mobile phones and some personal digital assistants.
- *Videoconferencing* uses a set of telecommunication technologies that allow two or more locations to interact simultaneously via two-way video and audio transmissions.

- *Virtual education* refers to instruction in which teachers and students are separated by time, space or both, and in which the teacher provides course content through a number of means, including course management applications, multimedia resources, the Internet and videoconferencing. Students receive the content and communicate with the teacher via the same technologies.
- *Voice over Internet Protocol (VoIP)* is the technology that enables phone calls over the Internet. The advantages include reduced costs on long-distance calls and voice mail that can be received as e-mail messages. Skype is one example of such a technology.
- *Web forums,* which evolved from pre-Web electronic bulletin boards, are the Web equivalent of a place where people can leave public messages that do such things as announce events, provide information or advertise things to buy or sell.
- A *webcast* is the broadcasting of audio or video content over the Internet. An example is a media file distributed over the Internet using streaming media technology. A webcast may be distributed either live or on demand.

Web 2.0[134]

Following a tradition in the naming of phases of software development, the term "Web 2.0" emerged in the wake of the 2001 bursting of the dot.com bubble to refer to a second generation of Web development. Web 2.0 does not refer to an update to any technical specifications of the Web but, rather, to changes in the ways software developers and endusers utilize the Web. Indeed, the term is more metaphorical than literal, since many of the technological components of Web 2.0 have existed since the early days of the Web, and the boundaries are fluid. Although the distinction is not clear and is often disputed, this second generation is roughly distinguished by using tools in a more interactive way, facilitating two-way communication and collaboration rather than the more static, one-way communication characteristic of Web 1.0.

At the time of this writing, Wikipedia defined Web 2.0 as "a perceived second generation of Web development and design that facilitates communication, secure information sharing, interoperability and collaboration on the World Wide Web. Web 2.0 concepts have led to the development and evolution of Web-based communities, hosted services and applications such as social-networking sites, video-sharing sites, wikis, blogs and folksonomies (taxonomies organically defined by people)." These features enable forms of participation and interaction that were not previously possible. Below, we define these and other technologies associated with Web 2.0, even though some of them actually predate this new phase.

The vocabulary can be confusing and is still being sorted out. When applied to business, the term Web 2.0 overlaps in usage with other terms, such as "Enterprise 2.0," "Social CRM" and "Social Business" (Morgan 2010b). Below is a list of the most frequently used terms that is meant to provide some clarity about these and many other terms:

- A *blog* (short for "Web log") is an online journal or diary hosted on a website. It may be maintained by an individual or a group that provides regular entries displayed in reverse-chronological order. A *microblog* is a blog post that has a restricted number of characters, such as Twitter (see below).

- *Blended learning* is a term that has evolved to describe the effort to systematically integrate different forms of learning as well as to combine and complement face-to-face instruction with the many other modes that are now possible.

- *The cloud:* The emerging next phase of the Internet features a combination of online resources (e.g., connections, software, services and servers) accessed over a network. This technology means that data is no longer stored on individual or organizational computers; instead, resources can be used and paid for only when needed, which also means that the maintenance of an IT infrastructure can be outsourced (Sankar and Bouchard 2009: 201). Terms such a "unified data center" and "unified cloud computing" are emerging. As one expert put it (Forrester, cited in ibid: 181): "A cloud is a pool of scalable, abstracted infrastructure that hosts end-use applications, billed by consumption."

- *Collective intelligence* refers to any system that attempts to tap the expertise of a group rather than of an individual so as to provide a service, produce a product or make decisions. Examples include collaborative publishing and common databases for sharing knowledge.

- *Crowdsourcing* describes the act of taking a task traditionally performed by an employee or contractor and outsourcing it to an undefined and generally large group of people or a community by making an open call for assistance. For example, the public may be invited to develop a new technology, carry out a design task or help capture, systematize or analyze large amounts of data.[135]

- *Enterprise 2.0* is a term popularized by Andrew McAfee in 2006,[136] used to describe social software used in "enterprise" (i.e., business) contexts (McAfee 2006b, 2006c). It includes social and networked modifications to company intranets and other classic software platforms used by large companies to organize their communication.

- *Internet forums* are a Web application for holding discussions and posting user-generated content. They are also commonly referred to in many other ways, including "Web forums," "newsgroups," "message boards," "discussion boards," "electronic discussion groups," "discussion forums," "bulletin boards," "fora" (the Latin plural of "forum") or simply "forums." Messages within these forums are displayed either in chronological order or as threaded discussions.

- *Mashups* are aggregations of content from different online sources that are used to create a new service. An example would be a program that pulls apartment listings from one site and displays them on a map provided by another service to show where the apartments are located (e.g., MyApartmentMap).

- *mLearning,* or "mobile learning," is a term describing learning via mobile phones or other portable technologies. It often consists of short modules that enable "just-in-time" learning or review, access to file systems or to Internet searches, and instant messaging.

- *Ning* is an example of an online community service that enables users to create their own social networks and join other social networks. Creators of networks can to some degree determine the site's appearance and functionality as well as which parts are public and private. Most networks include features such as: photos or videos; lists of network members, events, and groups within the network; and communication tools (e.g., forums or blogs).

- *Peer-to-peer networking* (sometimes called "P2P") is a technique for efficiently sharing files (music, video or text) either over the Internet or within a closed set of users. Unlike the traditional method of storing a file on one machine, which can become a bottleneck if many people try to access it at once, P2P distributes files across many machines, often those of the users themselves.

- *RSS* (Really Simple Syndication) is a family of "Web feed" formats used to publish frequently updated works (e.g., blog entries, news headlines, audio and video) in a standardized format. It allows people to get timely updates of information from favored websites or to aggregate information from many sites into one place.

- *Social bookmarking* enables users to share bookmarks of Web pages. These bookmarks are usually public, but they can also be private, shared only with specific people or groups, shared only within certain networks or shared within a combination of public and private domains. Services that support bookmarking for this purpose are Del.icio.us and StumbleUpon.

- *Social business* is defined by Morgan (2010b) as: "Enterprise that effectively collaborates both internally behind the firewall and externally with customers. ... (It) is built up on the concepts, strategies and integration of both social CRM and Enterprise 2.0. Evolving to a social business is a long-term strategic approach and, along with E2.0 and SCRM, includes things such as culture and corporate philosophy."

- *Social CRM* (or SCRM) is an abbreviation for "social customer relations management." It is "a strategy (oftentimes supported by technology) which allows organizations to make customers a focal point of how they do business." It usually means going beyond responding to customer e-mails and tweets to "fix(ing) the problems the customers are identifying and collaborating with your customers to help give them what they want." As such is it is "part of what being a social business is all about" (Morgan 2010a).

- *Social media* (sometimes known as "interactive sharing sites") allow users to post their own media content, to post comments about particular content and to vote to assess content. YouTube, Digg and Flickr are examples. Some people feel that this term is evolving to describe a new dimension of participation and is subsuming the concept of Web 2.0.[137]

- *Social networking* refers to systems that allow members of a specific site to learn about other members' skills, talents, knowledge or preferences as well as to communicate with them. Commercial examples include Facebook, MySpace and the aforementioned Ning. These tools are increasingly used by employers as a tool to communicate with their workforce (ILM 2009). One social-networking application that is growing rapidly is Twitter, which has features resembling a blog and a mobile-phone IM tool. (One blogger, Andrew McAfee, reports that it combines at least 17 familiar features into one function; cf. McAfee 2009b). Twitter allows users 140 characters for each posting (or "tweet") to say whatever they would like to say. Tweets appear on a public timeline and are displayed like a series of "microblogs." Registered users can subscribe to "follow" other users and thereby receive their tweets, and the others users can "follow" them in turn.

- *Unified communications* (UC) is the integration of real-time communication services (e.g., instant messaging [chat], presence information, telephony [including IP telephony], videoconferencing, call control and speech recognition) with non-real-time communication services (e.g., unified messaging [integrated voicemail, e-mail, SMS and fax]). UC is not a single product, but is rather a set of products that provides a consistent unified user interface and user experience across multiple devices and media types. UC also refers to a trend toward offering business process integration—that is, to simplify and integrate all forms of communications in order to optimize business processes and reduce response time, manage flows, and eliminate device and media dependencies.

- A *virtual world* is a computer-based simulated environment intended to be inhabited by its users and in which they interact via "avatars" (i.e., two- or three-dimensional graphical representations of the users). Users can manipulate elements of the modeled world and thereby experience a "virtual world," with rules based on the real world or some hybrid fantasy world. Second Life is an example of an online service enabling virtual worlds.

- *Webinars* are a specific type of "Web conference." Although the term originally described asynchronous group discussions and message boards, it now refers to "live" meetings. They are typically one-way—from the speaker to the audience—with limited audience interaction, as in a webcast. However, a webinar can be collaborative, including polling and question and answer sessions. In most cases, access may be supplemented by a telephone and/or Internet connection.

- *Web services* are layers of software technology that can be integrated with existing systems so as to make it easier for them to communicate with other systems and thereby automatically pass on information or conduct transactions. For example, a retailer and supplier might use Web services to communicate over the Internet and automatically update each other's inventory systems.

- A *wiki* is a Web page or collection of Web pages designed to enable anyone who accesses it to contribute or modify content by using a simplified markup language. This is an example of a so-called collective intelligence application. Wikis are often used to create collaborative websites and to support community websites. The collaborative encyclopedia Wikipedia is one of the best-known wiki applications. Wikis are also used in the business world to provide intranets and knowledge management systems.

Web 3.0 and Beyond

The semantic Web

According to Spivack's timeline, Web 3.0 has begun. A Google search for the term turns up hundreds of hits, many of them offering predictions of what is to come. Despite the lack of consensus about the best way to describe this future and its phases, some elements of what is coming next are already clear. For example, Web 3.0 will address the limitations of familiar search tools, such as Google. The service relies on keyword searches, a tool that will not be able to keep pace with the explosion of information made possible by the increased capacity for information creation brought about by the Web. Likewise, this type of search delivers the same information to everyone, failing to distinguish among users with different needs. More sophisticated approaches are already emerging. One example is Wolfram Alpha, a project launched in May 2009.

As one of the next steps, the person credited with inventing the Web, Tim Berners-Lee, calls for more "linked data," whereby easily made connections are established between or among raw data available on the Web (Berners-Lee 2009). He and others talk of the "semantic web," in which data are made more meaningful, either by putting them in the form of "metadata" using tools that include "tagging" (a Web 2.0 tool) and/or by creating smarter software that can describe relationships among different kinds of data. These innovations will enable more personalized responses to searches while also enabling the Internet to have the "intelligence" to identify tacit knowledge in users and make it explicit. Such an "implicit web" will enable a Web-based agent to track a person's pattern of Web usage and then to make inferences about the person's interests so as to provide suggestions without the person's having to make an explicit request for such information (van Allen 2009). Early examples of this are evident on Amazon.com and Netflix. The trend is away from a "push" model of marketing, in which companies are the initiators, to one in which consumers "pull" information relevant to their individual needs (Hagel III, Brown and Davison 2010; Siegel 2009).

It will be a while before the tantalizing possibilities of Web 3.0 will be realized, and it will be longer still before their implications for leadership are fully evident. The arrival of "the cloud" promises to provide the technological infrastructure for Web 3.0. Cloud computing has been called a "disruptive technology," and its entry has been compared to the development of the power grid as a stage in the evolution of the electric utility industry (Carr 2008). In this stage of transition, the ratio of costs and risks to reward for migration toward the cloud is probably too high for many organizations.

Experts at Cisco advise others to take it slowly, stating (Sankar and Bouchard 2009): "The cloud computing industry is in the gold rush stage: very young, many ideas, not enough general consensus and lots of companies (i.e., vendors). ... Because of the nature of the current state of the industry, it is better to start slow and follow the adoption curve." Nonetheless, the cloud seems to be popular among startups that have not yet invested in IT infrastructure as well as for companies that experience seasonal fluctuations in demand. Meanwhile, most organizations have yet to fully capitalize on the tools associated with Web 2.0.

Mobile phones

It is just now becoming clear that descriptions of Web 3.0 need to include the exponentially rising prevalence of mobile phones as the medium for accessing the Web. Meister and Willyerd observed that in 2010, more than 1.2 billion mobile phones were being produced each year and that they were benefitting from "unprecedented innovation." By 2020, they predict that (Meister and Willyerd 2010: 215): "Mobile phones and tablets will be the primary connection tool to the Internet for most people in the world" and that "your mobile device will become your office, your classroom and your concierge." Indeed, commenting on a 2009 McKinsey study on *HR Executive.com*, Matt Wilkinson, senior director of customer experience for SumTotal, predicted that Web 2.0 "will keep gaining momentum" (quoted in Starner 2009). Still, he believes the real breakout will occur as more and more Web 2.0 applications are ported to smart phones, BlackBerrys and other portable devices. "As Web 2.0 goes more outside of the personal computer," Wilkinson says, "you will see the level of engagement grow in more areas, especially in segments such as manufacturing and retail, where employees don't have easy access yet. We're going to see Web 2.0 really take off when that happens" (ibid.).

This trend has led to recent headlines like "The Web is Dead" (Anderson 2010). The basis for this proclamation (which was presumably overstated for dramatic effect) is that mobile phones shift access away from the open pattern of search that has thus far been characteristic of the Web and toward more channeled usage through "apps" (applications). In any case, mobile phone usage has greatly increased the popularity of "just-in-time" preparation and learning. Moreover, it promises to help overcome the digital divide by enabling Web access for people who do not possess a computer.

Index of Organizations

References

Anderson, Chris. *The Long Tail: Why the Future of Business Is Selling Less of More* (Revised and Updated). Redwood Shores, Calif.: Hyperion, 2008.

Anderson, Chris. "The Web Is Dead. Long Live the Internet." *Wired.com* September 2010. (www.wired.com/magazine/2010/08/ff_webrip/all/1, accessed Sept. 25,2010).

Anklam, Patti. *Net Work: A Practical Guide to Creating and Sustaining Networks at Work and in the World*. Amsterdam and Boston: Elsevier/Butterworth-Heinemann, 2007.

Argyris, Chris. *Increasing Leadership Effectiveness*. Wiley Series in Behavior. New York, N.Y.: Wiley, 1976.

Argyris, Chris, and Donald Schön. *Theory in Practice: Increasing Professional Effectiveness*. San Francisco, Calif.: Jossey-Bass, 1974.

Argyris, Chris, Robert Putnam and Diana McLain Smith. *Action Science*. The Jossey-Bass Social and Behavioral Science Series. San Francisco, Calif.: Jossey-Bass, 1985.

Armstrong, Charles. "Emergent Democracy." In *Open Government: Collaboration, Transparency and Participation in Practice*, edited by Daniel Lathrop and Laurel Ruma. Sebastopol, Calif.: O'Reilly Media, 2010: 167–176.

Barnes, Peter. *Capitalism 3.0: A Guide to Reclaiming the Commons*. San Francisco, Calif.: Berrett-Koehler Publishers, 2006.

Benkler, Yochai. *The Wealth of Networks: How Social Production Transforms Markets and Freedom*. New Haven, Conn.: Yale University Press, 2006.

Bennis, Warren, and Burt Nanus. *Leaders: The Strategies for Taking Charge*. New York, N.Y.: Harper and Row, 1985.

Berners-Lee, Tim. "Tim Berners-Lee on the Next Web." Video presentation for TED2009 in February 2009. (www.ted.com/index.php/talks/tim_berners_lee_on_the_next_web.html, accessed May 8, 2011).

Bernholz, Lucy, Edward Skloot and Barry Varela. *Disrupting Philanthropy: Technology and the Future of the Social Sector*. Durham, N.C.: Center for Strategic Philanthropy and Civil Society, Sanford School of Public Policy, Duke University, 2010.

Bodie, Grant. "How the Recession and Web 2.0 Is Changing Recruitment." *Real Business* April 24, 2009. (http://realbusiness.co.uk/news/how_the_recession_and_web_20_is_changing_recruitment, accessed July 30, 2010).

Boje, David M. "Network Leadership." New Mexico State University. Jan. 10, 2001. (http://cbae.nmsu.edu/~dboje/teaching/338/network_leadership.htm, accessed August 12, 2010).

Bornstein, David. *How to Change the World: Social Entrepreneurs and the Power of New Ideas.* New York, N.Y.: Oxford University Press, 2007.

Bradford, David L., and Allan R. Cohen. *Power Up: Transforming Organizations through Shared Leadership.* New York, N.Y.: J. Wiley, 1998.

Breck, Judy. *109 Ideas for Virtual Learning: How Open Content Will Help Close the Digital Divide.* Digital Learning Series. Lanham, Md.: Rowman & Littlefield, 2005.

Brotherton, David, and Cynthia Scheiderer. *Come On In. The Water's Fine. An Exploration of Web 2.0 Technology and Its Emerging Impact on Foundation Communications.* Brotherton Strategies. September 2008. (http://comnetwork.org/resources/brotherton_new_media_091608.pdf)

Brown, Judith, and David Isaacs. *The World Café: Shaping Our Futures through Conversations That Matter.* San Francisco, Calif.: Berret-Koehler Publishers, 2005.

Buhse, Willms, and Sören Stamer (eds.). *Enterprise 2.0: The Art of Letting Go.* Bloomington, Ind.: iUniverse, Inc., 2008.

Burns, James MacGregor. *Leadership.* New York, N.Y.: Harper & Row, 1978.

Carr, Nicholas. *The Big Switch: Rewiring the World, from Edison to Google.* New York, N.Y.: W. W. Norton & Company, 2008.

Carr, Nicholas. *The Shallows: What the Internet Is Doing to Our Brains.* New York, N.Y.: W.W. Norton & Company, 2010.

Center for Creative Leadership. *Catholic Healthcare Partners: Developing Next-Generation Leaders.* Greensboro, N.C., 2008.

Cross, Robert L., and Robert J. Thomas. *Driving Results through Social Networks: How Top Organizations Leverage Networks for Performance and Growth.* San Francisco, Calif.: Jossey-Bass, 2008.

Darling, Marilyn, Charles Parry and Joseph Moore. "Learning in the Thick of It." *Harvard Business Review* (83) 7/8: 84–92, July-August 2005. (http://hbr.org/2005/07/learning-in-the-thick-of-it/ar/1, accessed May 8, 2011).

"Deliberative Democracy: Ancient Athens online." *The Economist* May 6, 2010. (www.economist.com/node/16056622?story_id=E1_TGDVGGNN&CFID=166504202&CFTOKEN=58417672, accessed March 28, 2011).

DiJulio, Sarah, and Andrea Wood. *Online Tactics & Success: An Examination of the Obama for America New Media Campaign.* Wilburforce and Brainerd Foundations, 2009. (www.wilburforce.org/pdf/Online_Tactics_and_Success.pdf).

Dixon, Nancy M., Nate Allen, Tony Burgess, Pete Kilner and Steve Schweitzer. *CompanyCommand: Unleashing the Power of the Army Profession.* West Point, N.Y.: Center for the Advancement of Leadership Development, 2005.

Doyle, Michael, and David Strauss. *How to Make Meetings Work: The New Interaction Method.* New York, N.Y.: Wyden Books, 1976.

Drapeau, Mark. "Government 2.0: The Rise of the Goverati." *ReadWriteWeb.com.* Feb. 5, 2009. (www.readwriteweb.com/archives/government_20_rise_of_the_goverati.php, accessed June 5, 2010.)

Drath, Wilfred H., Cynthia D. McCauley, Charles J. Palus, Ellen Van Velsor, Patricia M.G. O'Connor and John B. McGuire. "Direction, Alignment, Commitment: Toward a More Integrative Ontology of Leadership." *Leadership Quarterly* 19: 635–653, 2008.

Eaves, David. "After the Collapse: Open Government and the Future of Civil Service." In *Open Government: Collaboration, Transparency and Participation in Practice,* edited by Daniel Lathrop and Laurel Ruma. Sebastopol, Calif.: O'Reilly Media, 2010: 139–152.

Eggers, William D. *Government 2.0: Using Technology to Improve Education, Cut Red Tape, Reduce Gridlock and Enhance Democracy.* Lanham, Md.: Rowman & Littlefield, 2007.

Erikson, Erik H. *Identity and the Life Cycle.* New York, N.Y.: W.W. Norton & Company, 1994.

Erickson, Tamara J. *What's Next, Gen X?: Keeping Up, Moving Ahead and Getting the Career You Want.* Cambridge, Mass.: Harvard Business Press, 2009.

Evans, Bob. "CIO at the CIA Recognized as IT leader, Change Agent." InformationWeek Global CIO Blog. April 29, 2009. (www.informationweek.com/blog/global-cio/229206851, accessed August 13, 2009).

Garekar, Bhagyashree. "Clinton Rides Web 2.0 Wave." *The Straits Times* February 27, 2010. (www.techxav.com/2010/02/27/clinton-rides-web-2-0-wave/, accessed March 28, 2011).

Gates, Bill. "A New Approach to Capitalism in the 21st Century." Speech delivered at the 2008 World Economic Forum in Davos, Switzerland. January 2008 (www.microsoft.com/presspass/exec/billg/speeches/2008/01-24wefdavos.mspx, accessed May 16, 2011).

Gegenhuber, Thomas. "Microsoft's e-Government Browser in Austria: Just a Good Service or a Play to Increase Market Share?" *Wikinomics.com* June 4, 2010. (http://www.wikinomics.com/blog/index.php/2010/06/04/microsofts-e-government-browser-in-austria-just-a-good-service-or-a-play-to-increase-market-share/, accessed May 23, 2010).

Gerencser, Mark, Reginald Van Lee, Fernando Napolitano and Christopher Kelly. *Megacommunities: How Leaders of Government, Business and Non-profits Can Tackle Today's Global Challenges Together.* New York, N.Y.: Palgrave Macmillan, 2009.

Gerstner, Louis V. *Who Says Elephants Can't Dance?: Leading a Great Enterprise through Dramatic Change.* New York, N.Y.: Harper Paperbacks, 2004.

Goleman, Daniel. *Emotional intelligence: Why It Can Matter More Than IQ* (10th ed.). New York, N.Y.: Bantam, 2006.

Govmonitor 2010. "Massachusetts Launches Real Time MBTA Bus Location Website." *Govmonitor* Sept. 10, 2010. (www.thegovmonitor.com/world_news/united_states/ massachusetts-launches-real-time-mbta-bus-location-website-38374.html, accessed Sept. 23, 2010).

Greenberg, Paul. "Time to Put a Stake in the Ground on Social CRM." Post on *Social-BizAgents* July 6, 2009. (http://socialbizagent.posterous.com/time-to-put-a-stake-in-the-ground-on-social-c, accessed May 8, 2011).

Griffin, Douglas, and Ralph Stacey (eds.). *Complexity and the Experience of Leading Organizations.* New York, N.Y.: Routledge, 2005.

Gunther, Marc. "Stonyfield Stirs up the Yogurt Market." *cnn.com* Jan. 4, 2008. (http:// money.cnn.com/2008/01/03/news/companies/gunther_yogurt.fortune/index.htm, accessed May 9, 2011).

Haeckel, Stephan H. *Adaptive Enterprise: Creating and Leading Sense-and-Respond Organizations.* Cambridge, Mass. Harvard Business Press, 1999.

Hagel III, John, John Seely Brown and Lang Davison. *The Power of Pull: How Small Moves, Smartly Made, Can Set Big Things in Motion.* New York, N.Y.: Basic Books, 2010.

Hamel, Gary. "The Facebook Generation vs. the Fortune 500." *Wall Street Journal* March 24, 2009. (http://blogs.wsj.com/management/2009/03/24/the-facebook-generation-vs-the-fortune-500/, accessed May 4, 2010).

Hanson, Rick, and Richard Mendius. *Buddha's Brain: The Practical Neuroscience of Happiness, Love and Wisdom.* Oakland, Calif.: New Harbinger, 2009.

Hart, Kim, and Megan Greenwell. "To Nonprofits Seeking Cash, Facebook App Isn't So Green." *washingtonpost.com* April 22, 2009. (www.washingtonpost.com/wp-dyn/ content/article/2009/04/21/AR2009042103786.html, accessed May 9, 2011).

Hassan, Zaid, and Mille Bojer. *The Change Lab Fieldbook.* Beverly, Mass.: Generon Consulting, 2005.

Hawken, Paul. *Blessed Unrest: How the Largest Movement in the World Came into Being and Why No One Saw It Coming.* New York, N.Y.: Viking, 2007.

Hawken, Paul, Amory Lovins and L. Hunter Lovins. *Natural Capitalism: Creating the Next Industrial Revolution.* Boston, Mass.: Back Bay Books, 2000.

Heifetz, Ronald A. *Leadership without Easy Answers.* Cambridge, Mass. Harvard Business Press, 1998.

Heifetz, Ronald A., and Donald L. Laurie. "The Work of Leadership." *Harvard Business Review* Dec. 2001: 131–140. (http://hbr.org/2001/12/the-work-of-leadership/ar/1, accessed May 8, 2011).

Heifetz, Ronald A., Marty Linsky and Alexander Grashow. *The Practice of Adaptive Leadership: Tools and Tactics for Changing Your Organization and the World.* Cambridge, Mass.: Harvard Business Press, 2009.

Hernandez, Manny. "What Is TuDiabetes?" Post on *tudiabetes.org* Aug. 7, 2008. (www. tudiabetes.org/notes/index/show?noteKey=What_is_TuDiabetes%3F, accessed May 9, 2011).

Hernandez, Manny. "Introducing TuAnalyze: Why Mapping Diabetes Data Matters." Blog post on *askmanny.com* May 24, 2010. (http://askmanny.com/2010/05/introducing-tuanalyze-why-mapping-diabetes-data-matters/, accessed May 9, 2011).

Holland, John H. *Hidden Order: How Adaptation Builds Complexity*. New York, N.Y.: Basic Books, 1996.

Howe, Jeff. *Crowdsourcing: Why the Power of the Crowd is Driving the Future of Business* (unedited edition). New York, N.Y.: Crown Business, 2009.

Hubbard, Barbara. *Investing in Leadership (Vol. 1): A Grantmaker's Framework for Understanding Nonprofit Leadership Development*. Washington, D.C.: Grantmakers for Effective Organizations, 2005.

Hurley, Thomas J. *Leadership in Networked, Emergent Systems*. Unpublished Power-point presentation at the Shambhala Institute for Authentic Leadership, June 2007.

Iansiti, Marco, and Roy Levien. *The Keystone Advantage: What the New Dynamics of Business Ecosystems Mean for Strategy, Innovation and Sustainability*. Boston, Mass.: Harvard Business School Publishing, 2004.

IBM Redguide. *Enterprise Innovation and Technology Adoption Guide*. 2009. (www.redbooks.ibm.com/abstracts/redp4520.html, accessed May 8, 2011).

ILM (Institute of Leadership and Management). "Firms 'Embracing Web 2.0'." Posting on company website April 22, 2009. (www.i-l-m.com/members/2363.aspx?articleid=19133419&, accessed May 8, 2011).

Jackson, Joab. "Intellipedia Suffers Midlife Crisis." *GCN (Government Computer News)* Feb. 18, 2009. (http://gcn.com/articles/2009/02/18/intellipedia.aspx, accessed May 5, 2010).

Jellinek, Dan. "Special Focus – Participatory Budgeting, Part 2: Sympathy for the Devil?" Blog post on *headstar.com* Nov. 4, 2008. (www.headstar.com/egblive/?p=172, accessed May 9, 2011).

Joiner, William B., and Stephan A. Josephs. *Leadership Agility: Five Levels of Mastery for Anticipating and Initiating Change*. San Francisco, Calif.: Jossey-Bass, 2006.

Kahane, Adam. *Power and Love: A Theory and Practice of Social Change*. San Francisco, Calif.: Berret-Koehler Publishers, 2010.

Kanter, Beth. "Amanda Rose, Twestival Leader, Reflects on Twestival." Post on *Beth's Blog* Feb. 17, 2009a. (http://beth.typepad.com/beths_blog/2009/02/amanda-rose-twe.html, accessed May 11, 2011).

Kanter, Beth. "The Crumbling of Nonprofit Arts Organizations: What Models Will Rise from the Ashes?" Post on *Beth's Blog* March 19, 2009b. (http://beth.typepad.com/beths_blog/2009/03/the-crumbling-of-nonprofit-arts-organizations-what-models-will-rise-from-the-ashes.html, accessed May 11, 2011).

Kanter, Beth. "Smithsonian: Crowdsourcing An Institution's Vision on YouTube." Post on *Beth's Blog* May 25, 2009c. (http://beth.typepad.com/beths_blog/2009/05/smithsonian-crowdsourcing-an-institutions-vision-on-youtube.html, accessed May 11, 2011).

Kanter, Beth. "The Red Cross Created A Social Culture With Listening: Now, A Policy and Operational Handbook To Scale." Post on *Beth's Blog* July 6, 2009d. (http://beth.typepad.com/beths_blog/2009/07/red-cross-social-media-strategypolicy-handbook-an-excellent-model.html, accessed May 11, 2011).

Kanter, Beth. "Can A Public Nonprofit Dashboard Be Your Organization's Best Friend or Worst Enemy?" Post on *Beth's Blog* Nov. 3, 2009e. (http://beth.typepad.com/beths_blog/2009/11/is-a-publically-shared-dashboard-your-nonprofits-best-friend.html, accessed May 11, 2011).

Kanter, Beth. "A Few Reflections from SXSW Crowdsourcing Panel." Post on *Beth's Blog* March 16, 2010a. (http://beth.typepad.com/beths_blog/2010/03/a-few-reflections-from-sxsw-crowdsourcing-panel.html, accessed May 11, 2011).

Kanter, Beth. "The Social Media Trainer's Bag of Tricks." Post on *Beth's Blog* Jan. 26, 2010b. (http://socialmedia-for-trainers.wikispaces.com/, accessed May 11, 2011).

Kanter, Beth. "Should CEOs and Executive Directors Use Social Media?" Post on *Beth's Blog* Feb. 9, 2010c. (http://beth.typepad.com/beths_blog/2010/02/should-ceos-and-executive-directors-use-social-media.html, accessed May 11, 2011).

Kanter, Beth. "Got Social Media Policy?" Post on *Beth's Blog* March 8, 2010d. (http://beth.typepad.com/beths_blog/2010/03/got-social-media-policy-.html, accessed May 11, 2011).

Kanter, Beth. "What is Lethal Generosity?" Post on *Beth's Blog* April 21, 2010e. (http://beth.typepad.com/beths_blog/2010/04/lethal-generosity-defined.html, accessed May 11, 2011).

Kanter, Beth. "Wisdom 2.0: Living Consciously in a Connected World." Post on *Beth's Blog* May 1, 2010f. (http://beth.typepad.com/beths_blog/2010/05/wisdom-20-balance-of-technology-use-is-not-an-onoff-switch.html, accessed May 11, 2011).

Kanter, Beth. "Microsoft's Networked Approach To Accelerating Social Change Through Technology." Post on *Beth's Blog* June 1, 2010g. (www.bethkanter.org/unlimited-potential/, accessed May 11, 2011).

Kanter, Beth. "How Many Free Agents Does It Take To Change A Nonprofit Fortress?" Post on *Beth's Blog* June 4, 2010h. (www.bethkanter.org/lightbulb-fortress-freeagent/, accessed May 11, 2011).

Kanter, Beth. "Networked Nonprofits Deconstruct Social Media Fears." Post on *Beth's Blog* Aug. 8, 2010i. (www.bethkanter.org/wallow-in-fear/, accessed May 11, 2011).

Kanter, Beth. "Celebrating Free Agents: Mark Horvath Launches WeAreVisible." Post on *Beth's Blog* Sept. 14, 2010i. (www.bethkanter.org/wearevisible/?utm_source=feedburner&utm_medium=email&utm_campaign=Feed%3A+bethblog+%28Beth%27s+Blog%29, accessed May 11, 2011).

Kanter, Beth. "A Free Agent and Nonprofits Working Together To End Homelessness." Post on *Beth's Blog* Oct. 5, 2010k. (www.bethkanter.org/invisible-people/, accessed May 11, 2011).

Kanter, Beth, and Allison Fine. *The Networked Nonprofit: Connecting with Social Media to Drive Change.* San Francisco, Calif.: Jossey-Bass, 2010.

Kasper, Gabriel, and Diana Scearce. "Working Wikily: How Networks Are Changing Social Change." Monitor Institute and the David and Lucile Packard Foundation, 2008. (www.workingwikily.net/Working_Wikily.pdf).

Kearns, Marty. "Network-Centric Advocacy." Green Media Toolshed, 2005. (http://activist.blogs.com/networkcentricadvocacypaper.pdf).

Kearns, Marty, and Karyn Showalter. "Network-Centric Advocacy: Like the Dog that Catches the Car …" Netcentric Campaigns, n.d. (www.connectusfund.org/resources/network-centric-advocacy-dog-catches-car, accessed May 16, 2011).

Kegan, Robert. *The Evolving Self: Problem and Process in Human Development.* Cambridge, Mass.: Harvard University Press, 1994.

Kegan, Robert. *In Over Our Heads: The Mental Demands of Modern Life.* Cambridge, Mass.: Harvard University Press, 1998.

Kegan, Robert, and Lisa Laskow Lahey. *Immunity to Change: How to Overcome It and Unlock the Potential in Yourself and Your Organization.* Boston, Mass.: Harvard Business School Publishing, 2009.

Kelly, Kevin. *Out of Control: The Rise of Neo-Biological Civilization.* Boston, Mass.: Addison Wesley, 1994.

Kelly, Kevin. *What Technology Wants.* New York, N.Y.: Viking Adult, 2010.

Kelty, Christopher M. *Two Bits: The Cultural Significance of Free Software.* Durham, N.C.: Duke University Press Books, 2008.

Kobielus, James G. "Mighty Mashups: Do-It-Yourself Business Intelligence for the New Economy." Forrester Research. July 23, 2009. (www.forrester.com/rb/search/results.jsp?N=0+11889, accessed August 19, 2010.)

Kruse, Peter. "The Power of the Network." *catboant.com* Jan. 11, 2010. (www.catboant.com/category/conversations/peter-kruse/, accessed May 9, 2011).

Kruse, Peter. "The Network Is Challenging Us." *we-magazine.net* n.d. (www.we-magazine.net/we-volume-03/the-network-is-challenging-us/, accessed May 9, 2011).

Lagace, Martha. "Nonprofit Networking: The New Way to Grow." *Harvard Business School Working Knowledge* May 16, 2005. (http://hbswk.hbs.edu/archive/4801.html, accessed May 9, 2011).

Lancaster, Lynne C., and David Stillman. *When Generations Collide: How to Solve the Generational Puzzle at Work.* New York, N.Y.: HarperCollins, 2002.

Lanier, Jaron. "Digital Maoism: The Hazards of the New Online Collectivism." *Edge: The Third Culture.* May 30, 2006. (www.edge.org/3rd_culture/lanier06/lanier06_index.html, accessed April 22, 2010.)

Lanier, Jaron. *You Are Not a Gadget: A Manifesto.* New York, N.Y.: Knopf, 2010.

La Piana Consulting. "Convergence: How Five Trends Will Reshape the Social Sector." Nov. 2009. (www.lapiana.org/downloads/Convergence_Report_2009.pdf).

Lathrop, Daniel, and Laurel Ruma (eds.). *Open Government: Collaboration, Transparency and Participation in Practice.* Sebastapol, Calif.: O'Reilly Media, 2010.

Lepsinger, Richard, and Darleen DeRosa. *Virtual Team Success: A Practical Guide for Working and Leading from a Distance.* San Francisco, Calif.: John Wiley & Sons, 2010.

Lessig, Lawrence. *The Future of Ideas: The Fate of the Commons in a Connected World.* New York, N.Y.: Random House, 2001.

Levine, Rick, Christopher Locke, Doc Searls and David Weinberger. *The Cluetrain Manifesto (*10th Anniversary Edition). New York, NY: Basic Books, 2009.

Li, Charlene. *Open Leadership: How Social Technology Can Transform the Way You Lead.* San Francisco, Calif.: Jossey-Bass, 2010.

Li, Charlene, and Josh Bernoff. *Groundswell: Winning in a World Transformed by Social Technologies.* Boston, Mass.: Harvard Business School Publishing, 2008.

Lichtenstein, Jesse. "Digital Diplomacy." *NYTimes.com* July 16, 2010. (www.nytimes.com/2010/07/18/magazine/18web2-0-t.html, accessed May 9, 2011).

Loevinger, Jane. *Ego Development: Conceptions and Theories.* San Francisco, Calif.: Jossey-Bass, 1976.

Luckey, Amy, Suki O'Kane and Eric Nee. "Web 2.0 for Grantmakers: Using New Internet Technology to Increase Grantmaking Effectiveness." Handout for presentation at "Ideas to Action" conference sponsored by Grantmakers for Effective Organizations in San Francisco, Calif., March 10–12, 2008. (www.geofunders.org/document.aspx?oid=a0660000003YTYA, accessed May 16, 2011.)

Mackey, John, Milton Friedman and T.J. Rodgers. "Rethinking the Social Responsibility of Business." *reason.com* October 2005. (http://reason.com/archives/2005/10/01/rethinking-the-social-responsi, accessed May 3, 2011).

Malone, Phil. An Evaluation of Private Foundation Copyright Licensing Policies, Practices and Opportunities. Cambridge, Mass.: Berkman Center for Internet & Society, 2009. (http://cyber.law.harvard.edu/publications/2009/Open_Content_Licensing_for_Foundations, accessed May 3, 2011).

Marshall, Alex. "Collaborative Public Policy-Making, the Freiburg Way." *Wikinomics.com* February 16, 2009. (www.wikinomics.com/blog/index.php/2009/02/16/collaborative-public-policy-making-the-freiburg-way/, accessed May 10, 2011).

Martin, Dan. "What Can We Learn from Eurostar's Social Media Shortcomings?" *Mycustomer.com* July 7, 2010. (www.mycustomer.com/topic/eurostar, accessed May 10, 2011).

McAfee, Andrew. "Enterprise 2.0: The Dawn of Emergent Collaboration." *MIT Sloan Management Review* April 12, 2006a. (http://sloanreview.mit.edu/the-magazine/2006-spring/47306/enterprise-the-dawn-of-emergent-collaboration/, accessed May 10, 2011).

McAfee, Andrew. "Enterprise 2.0 vs. SOA." Blog post on *andrewmcafee.org.* May 29, 2006b. (http://andrewmcafee.org/2006/05/enterprise_20_vs_soa/, accessed May 10, 2011).

McAfee, Andrew.. "Enterprise 2.0, Version 2.0." Blog post on *andrewmcafee.org* May 27, 2006c. (http://andrewmcafee.org/2006/05/enterprise_20_version_20/, accessed Oct. 1, 2010).

McAfee, Andrew. *Enterprise 2.0: New Collaborative Tools for Your Organization's Toughest Challenges.* Boston, Mass.: Harvard Business School Publishing, 2009a.

McAfee, Andrew. "17 Things We Used to Do." Blog post on *Andrew McAfee's Blog*. April 20, 2009b. (http://andrewmcafee.org/2009/04/17-things-we-used-to-do/, accessed May 16, 2011).

McCauley, Cynthia D., and Stéphane Brutus. *Management Development through Job Experiences: An Annotated Bibliography*. Greensboro, N.C.: Center for Creative Leadership, 1998. (www.ccl.org/leadership/pdf/research/MgmtDevelopmentBib.pdf).

McCauley, Cynthia D., and Ellen Van Velsor (eds.). *Center for Creative Leadership Handbook of Leadership Development* (2nd ed.). San Francisco, Calif.: Jossey-Bass, 2004.

McGuire, John B., and Gary Rhodes. *Transforming Your Leadership Culture*. San Francisco, Calif.: Jossey-Bass, 2009.

McIntosh, Steve. *Integral Consciousness and the Future of Evolution*. St. Paul, Minn.: Paragon House, 2007.

McKinsey & Co. 2007. "How businesses are using Web 2.0: A McKinsey Global Survey." *McKinsey Quarterly* March 2007. (https://www.mckinseyquarterly.com/How_businesses_are_using_Web_20_A_McKinsey_Global_Survey_1913, accessed May 15, 2009).

Meister, Jeanne C., and Karie Willyerd. *The 2020 Workplace: How Innovative Companies Attract, Develop and Keep Tomorrow's Employees Today*. New York, N.Y.: HarperBusiness, 2010.

Mezey, Matthew. "Open leadership + Enterprise 2.0: The Practices That Can Make Them Real." Blog post on *Library & Information Update* May 10, 2010. (http://communities.cilip.org.uk/blogs/update/archive/2010/05/10/open-leadership-enterprise-2-0-the-practices-that-can-make-them-real.aspx, accessed May 11, 2011).

Michalski, Jerry. "Don Tapscott – strategist, author of "Wikinomics." Video Interview. *Fastforward* Feb. 8, 2008. (www.fastforwardblog.com/2008/02/18/don-tapscott-strategist-author-of-wikinomics/, accessed May 8, 2011).

Mintzberg, Henry. *The Rise and Fall of Strategic Planning*. New York; N.Y.: Free Press, 1994.

Monitor Institute and David and Lucile Packard Foundation 2007. "What Networks Do and Why They Matter Now." Fall 2007. (www.virtualsocialmedia.com/how-social-media-will-work-for-your-business-social-media-marketing/, accessed July 12, 2010).

Morgan, Jacob. "What Is the Real Value of Social CRM (SCRM)?" Blog Post on *Jacob Morgan* May 14, 2010a. (www.jmorganmarketing.com/real-value-social-crm-scrm/, accessed May 12, 2011).

Morgan, Jacob. "Social Media vs. Social CRM vs Social Business vs Enterprise 2.0." Blog post on *Jacob Morgan May 19, 2010b. (*www.jmorganmarketing.com/social-media-social-crm-social-business-enterprise-2-0/, *accessed May 12, 2011)*.

Mulholland, Andy, Chris S. Thomas and Paul Kurchina. *Mashup Corporations: The End of Business as Usual*. New York, N.Y.: Evolved Technologist, 2007.

N+1 Editorial Staff. "Webism: The Social Movement." *N+1* 9 (Spring 2010).

Northouse, Peter G. *Leadership: Theory and Practice* (4th ed.). Thousand Oaks, Calif.: Sage Publications, 2006.

Noveck, Beth Simone. "Wiki-Government: How Open Source Technology Can Make Government Decision-Making More Expert and More Democratic." *Democracy (A Journal of Ideas)* (7). Winter 2008. (www.democracyjournal.org/7/6570.php, accessed May 11, 2011).

Noveck, Beth Simone. *Wiki Government: How Technology Can Make Government Better, Democracy Stronger and Citizens More Powerful.* Washington, D.C.: Brookings Institution Press, 2009.

Noveck, Beth Simone. "Open Government Plans: A Tour of the Horizon" (The White House Open Government Initiative). *whitehouse.gov* April 8, 2010. (www.whitehouse.gov/blog/2010/04/08/open-government-plans-a-tour-horizon, accessed May 11, 2011)

Oshry, Barry. *Leading Systems: Lessons from the Power Lab.* San Francisco, Calif.: Berrett-Koehler Publishers, 1999.

Oshry, Barry. *Seeing Systems: Unlocking the Mysteries of Organizational Life.* San Francisco, Calif.: Berrett-Koehler Publishers, 2007.

Owen, Harrison. *Open Space Technology: A User's Guide.* San Francisco, Calif.: Berrett-Koehler Publishers, 2008.

Petzinger, Thomas. *The New Pioneers: The Men and Women Who Are Transforming the Workplace and Marketplace.* New York, N.Y.: Simon & Schuster, 1999.

Pople, Richard. "Government Takes Web 2.0 with a Web 1.0 Mindset." *Federal Computer Week* April 15, 2009. (http://fcw.com/Articles/2009/04/20/Pople-Comment.aspx, accessed March 22, 2011.)

Powers, William. *Hamlet's Blackberry: A Practical Philosophy for Building a Good Life in the Digital Age.* New York, N.Y.: Harper, 2010.

Rheingold, Howard. *Smart Mobs: The Next Social Revolution.* Cambridge, Mass.: Basic Books, 2003.

Rheingold, Howard. Best Business Books: The Future. *Strategy+Business* Winter 2006.

Saint-Onge, Hubert. *Collaborative Knowledge and Competitive Advantage.* New Paradigm Big Idea Series. 2005.

Sankar, Krishna, and Susan A. Bouchard. *Enterprise Web 2.0 Fundamentals.* Indianapolis, Ind.: Cisco Press, 2009.

Scharmer, C. Otto. *Theory U: Leading from the Future as It Emerges.* San Francisco, Calif.: Berrett-Koehler Publishers, 2009.

Schein, Edgar H. "The Anxiety of Learning." *Harvard Business Review* (80) 3: 100–107, 2002.

Schein, Edgar H. *DEC is Dead, Long Live DEC: The Lasting Legacy of Digital Equipment Corporation.* San Francisco, Calif.: Berret-Koehler Publishers, 2003.

Schein, Edgar H. *Helping: How to Offer, Give and Receive Help.* San Francisco, Calif.: Berrett-Koehler Publishers, 2009.

Schein, Edgar H. *Organizational Culture and Leadership.* San Francisco, Calif.: Jossey-Bass, 2010.

Scherer, Michael. "Salon Person of the Year: S.R. Sidarth." *Salon* Dec. 16, 2006. (www.salon.com/news/opinion/feature/2006/12/16/sidarth/print.html, accessed May 3, 2011).

Schooley, Claire. *Social Learning Tools: Changing the Learning Process*. Forrester Research, Inc. April 23, 2009. Slides from a webinar presentation.

Schwartz, Jon. "Flash Activists Use Social Media to Drum Up Support." *USA Today* May 5, 2009. (www.usatoday.com/tech/news/2009-05-05-flash-activists-protests_N.htm, accessed May 9, 2011).

Schwartz, Peter. *The Art of the Long View: Planning for the Future in an Uncertain World*. New York, N.Y.: Currency Doubleday, 1996.

Schwartz, Peter. *The Skilled Facilitator: A Comprehensive Resource for Consultants, Facilitators, Managers, Trainers and Coaches*. San Francisco, Calif.: Jossey-Bass, 2002.

Schwartz, Richard C. *Internal Family Systems Therapy*. New York, N.Y.:The Guilford Press, 1997.

Senge, Peter M. *The Fifth Discipline: The Art & Practice of the Learning Organization* (revised edition). New York, N.Y.: Broadway Business, 2006.

Senge, Peter M., C. Otto Scharmer, Joseph Jaworski and Betty Sue Flowers. *Presence: An Exploration of Profound Change in People, Organizations and Society*. New York, N.Y.: Broadway Business, 2005.

Senge, Peter M., Bryan Smith, Nina Kruschwitz, Joe Laur and Sara Schley. *The Necessary Revolution: Working Together to Create a Sustainable World*. New York, N.Y.: Broadway Books, 2010.

Shapira, Ian. "What Comes Next After Generation X?" *The Washington Post* July 6, 2008: C01. (www.washingtonpost.com/wp-dyn/content/article/2008/07/05/AR2008070501599.html, accessed May 8, 2011).

Shirky, Clay. *Here Comes Everybody: The Power of Organizing Without Organizations*. New York, N.Y.: Penguin Press, 2008.

Shirky, Clay. *Cognitive Surplus: Creativity and Generosity in a Connected Age*. New York, N.Y.: Penguin Press, 2010.

Shuen, Amy. *Web 2.0: A Strategy Guide*. Sebastopol, Calif.: O'Reilly Media, 2008.

Siegel, Daniel J. *The Mindful Brain: Reflection and Attunement in the Cultivation of Well-Being*. New York, N.Y.: W.W. Norton, 2007.

Singh, Lisa. "Web 2.0 and the Military: Booz Allen's Art Fritzson on What's Next." *ExecutiveBiz* May 4, 2009. (http://blog.executivebiz.com/web-20-adoption-within-the-military/1756, accessed April 12, 2010).

Slaughter, Anne-Marie. "America's Edge: Power in the Networked Century." *Foreign Affairs* (Jan./Feb. 2009). (www.foreignaffairs.com/articles/63722/anne-marie-slaughter/americas-edge, accessed May 11, 2011).

Snowden, David. "The Dogmas of the Quiet Past." *Cognitive Edge* Oct. 27, 2008. (www.cognitive-edge.com/blogs/dave/2008/10/the_dogmas_of_the_quiet_past_1.php, accessed May 26, 2010.)

Snowden, David. "Everything Is Fragmented—The Core Principles." *KMWorld* Jan. 2, 2009. (www.kmworld.com/Articles/News/News-Analysis/Everything-is-fragmented–The-core-principles–52016.aspx, accessed May 5, 2011).

Spivack, Nova. "How The WebOS Evolves?" *Novaspivack.com* February 09, 2007. (http://novaspivack.typepad.com/nova_spivacks_weblog/2007/02/steps_towards_a.html, accessed June, 2009.)

Stacey, Ralph D. *Complexity and Organizational Reality: Uncertainty and the Need to Rethink Management after the Collapse of Investment Capitalism* (2nd ed.). New York, N.Y.: Routledge, 2010.

Starner, Tom. "Measuring the Value of Web 2.0." *hreonline.com* Oct. 15, 2009. (http://www.hreonline.com/HRE/story.jsp?storyId=270696740, accessed on May 11, 2011).

Surowiecki, James. *The Wisdom Of Crowds: Why the Many Are Smarter than the Few and How Collective Wisdom Shapes Business, Economies, Societies and Nations.* New York, N.Y.: Doubleday, 2004.

Tapscott, Don. "Winning with the Enterprise 2.0." In *Enterprise 2.0: The Art of Letting Go*, edited by Willms Buhse and Sören Stamer. Bloomington, Ind.: iUniverse, 2008a: 98–120.

Tapscott, Don. "Success in the Second Wave of the Internet." *imakenews.com* May 2008b. (www.imakenews.com/ciscotcc/e_article001086217.cfm?x=bcC90Vp,b7TGJLBr,w, accessed May 11, 2011).

Tapscott, Don. *Growing Up Digital: How the Net Generation is Changing Your World.* New York, N.Y.: McGraw-Hill, 2008c.

Tapscott, Don, and Richard Goodwin. "Success in the Second Wave of the Internet: How Wikinomics Creates Opportunities for Cisco Partners." *New Paradigm* March 2008. (http://newsroom.cisco.com/dlls/2008/ekits/SuccessOfTheSecondWave_apr0208.pdf)

Tapscott, Don, and Anthony D. Williams. *Wikinomics: How Mass Collaboration Changes Everything.* New York, N.Y.: Portfolio Hardcover, 2006.

Tapscott, Don, and Anthony D. Williams. *Macrowikinomics: Rebooting Business and the World.* New York, N.Y.: Portfolio Hardcover, 2010.

Thaler, Richard H., and Cass R. Sunstein. *Nudge: Improving Decisions about Health, Wealth and Happiness.* New York, N.Y.: Penguin (Non-Classics), 2009.

Thompson, Clive. "Open Source Spying." *New York Times* December 3, 2006. (http://www.nytimes.com/2006/12/03/magazine/03intelligence.html, accessed May 9, 2011).

Ticoll, David, and Phil Hood. *The Dancing Penguin: Harnessing Self-Organization for Competitive Advantage.* T&CA Big Ideas, IT&CA Program. New York, N.Y.: New Paradigm, 2005.

Torbert, William R. *Creating a Community of Inquiry: Conflict, Collaboration, Transformation.* New York, N.Y.: John Wiley & Sons, 1976.

Torbert, William R., with Susanne Cook-Greuter, Dalmar Fisher, Erica Foldy, Alain Gauthier, Jackie Keeley, David Rooke, Sara Ross, Catherine Royce, Jenny Rudolph,

Steve Taylor and Mariana Tran. *Action Inquiry: The Secret of Timely and Transforming Leadership.* San Francisco, Calif.: Berrett-Koehler Publishers, 2004.

Tozzi, John. "Gov 2.0: The Next Internet Boom." *BusinessWeek.com* May 27, 2010. (www.businessweek.com/smallbiz/content/may2010/sb20100526_721134.htm, accessed May 8, 2011).

U.S. Department of Defense. *Network-Centric Warfare.* Report to Congress, July 27, 2001. (http://cio-nii.defense.gov/docs/pt2_ncw_main.pdf).

van Allen, Philip. "The Implicit Web: A New Trend." Post on *Philip van Allen.* April 27, 2009. (www.philvanallen.com/2009/04/the-implicit-web-a-new-trend, accessed April 20, 2010.)

van Slyke, David M., and Robert W. Alexander. "Public Service Leadership: Opportunities for Clarity and Coherence." *The American Review of Public Administration* (36) 4: 362–373, 2006.

Vance, Ashlee. "3-D Printing Spurs a Manufacturing Revolution." *nytimes.com* Sept. 14, 2010. (www.nytimes.com/2010/09/14/technology/14print.html?scp=1&sq=ashlee%20vance%20sept.%2014,%202010&st=cse, accessed May 11, 2011).

Vega, Tanzina. "New Web Code Draws Concern Over Privacy Risks." *nytimes.com* Oct. 11, 2010. (www.nytimes.com/2010/10/11/business/media/11privacy.html, accessed May 11, 2011).

Velsor, Ellen Van, Cynthia D. McCauley and Marian N. Ruderman. *The Center for Creative Leadership Handbook of Leadership Development* (3rd editon). San Francisco, Calif.: Jossey-Bass, 2010.

Volckmann, Russ. "Integral Leadership Theory." In *Political and Civic Leadership: A Reference Handbook*, Vol. 1, edited by Richard A. Couto. Los Angeles, Calif.: Sage, 2010: 121–127.

Watson, Tom. *CauseWired: Plugging In, Getting Involved, Changing the World.* Hoboken, N.J.: Wiley, 2009.

Weeks, Linton. "The Extraordinaries: Will Microvolunteering Work?" *npr.org* July 1, 2009. (www.npr.org/templates/story/story.php?storyId=106118736, accessed May 11, 2011).

Weinberger, David. *The Cluetrain Manifesto.* New York, N.Y.: Basic Books, 2000.

Weinberger, David. "The Risk of Control." In *Enterprise 2.0: The Art of Letting Go*, edited by Willms Buhse and Sören Stamer. Bloomington, Ind.: iUniverse, 2008: 66–73.

Wenger, Etienne, Richard McDermott and William Snyder. *Cultivating Communities of Practice: A Guide to Managing Knowledge.* Boston, Mass.: Harvard Business School Press, 2002.

Wheatley, Margaret. *Leadership and the New Science: Learning about Organization from an Orderly Universe.* San Francisco, Calif.: Berret-Koehler Publishers, 1992.

Wilber, Ken. *A Brief History of Everything.* Boston, Mass.: Shambhala, 2000.

Wilber, Ken. *A Theory of Everything: An Integral Vision for Business, Politics, Science and Spirituality.* Boston, Mass.: Shambhala, 2001.

Williams, Anthony D. *The New Innovation: Rethinking Intellectual Property for the ONE.* New York, N.Y.: New Paradigm, 2005.

Williams, Anthony D. (2008a). "Government 2.0: Wikinomics and the Challenge to Government." *Canadian Government Executive* May 30, 2008a. (http://cge.itincanada.ca/index.php?cid=314&id=6765&np=5, accessed March 28, 2011).

Williams, Anthony D. "50,000 Estonians Clean Up Their Country in One Day." Blog post on *anthonydwilliams.com* May 28, 2008b. (http://anthonydwilliams.com/2008/05/28/50000-estonians-clean-up-their-country-in-one-day/, accessed March 28, 2011).

Williams, Anthony D. "Open Forum Europe: The Openness Imperative." Blog post on *anthonydwilliams.com* April 6, 2009. (http://anthonydwilliams.com/2009/04/06/open-forum-europe-the-openness-imperative/, accessed May 11, 2011).

Williams, Anthony D. "Wikinomics and the Era of Openness: European Innovation at the Crossroads." Blog post on *anthonydwilliams.com* March 10, 2010a. (http://anthonydwilliams.com/2010/03/10/wikinomics-and-the-era-of-openness-european-nnovation-at-the-crossroads/, accessed May 11, 2011).

Williams, Anthony D. "Cognitive Surpluses and Deficits." Blog post on *anthonydwilliams.com* Aug. 23, 2010b. (http://anthonydwilliams.com/2010/08/23/cognitive-surpluses-and-deficits/, accessed May 11, 2011).

Williams, Anthony D., and Heidi Hay. "Digital-Era Policy-Making." Digital4Sight. 2000. (http://anthonydwilliams.com/wp-content/uploads/2006/08/Digital%20Era%20Policy%20Making.pdf).

Winer, Laurie. "Born to Check Mail." A book review of William Power's *Hamlet's Blackberry. New York Times* July 16, 2010. (www.nytimes.com/2010/07/18/books/review/Winer-t.html, accessed May 8, 2011).

Yunus, Muhammad. "Halving Poverty by 2015: We Can Actually Make It Happen." Commonwealth Lecture 2003. March 11, 2003. (www.grameen-info.org/index.php?option=com_content&task=view&id=220&Itemid=172, accessed May 11, 2011).

Endnotes

1 Thanks to Thomas Petzinger (Petzinger 1999) for the inspiration for this quote.

2 Patti Anklam, Patricia Arnold, Debra Beck, Willms Buhse, Stephen Cummings, Marilyn Darling, Naava Frank, Thomas Gegenhuber, Tom Glaisyer, Nauman Haque, Peter Hollands, Joitske Hulsebosch, Elmar Husmann, Jonathan Imme, Beth Kanter, Brenda Kaulback, Eugene Eric Kim, Alice MacGillivray, Barbara McDonald, Nicco Mele, Christina Merl, Matthew Mezey, Jerry Michalski, Lee Rainee, Eric-Marie Picard, Allyson Reaves, Claire Reinelt, Ulrike Reinhard, Andre Richier, Frank Roebers, Sebastian Schmidt, John David Smith, Bill Torbert, Leonard Waks, David Weinberger, Nancy White, Ole Wintermann and Abby Yanow.

3 We find the "integral" paradigm to be a powerful one, and it has attracted an ecosystem of theorists and practitioners (McIntosh 2007). However, a good book on leadership from this perspective has yet to be written. The best source on this topic is the online journal *Integral Leadership Review*, edited by Russ Volckman.

4 N+1 editorial staff, 2010.

5 Anthony Williams provides a balanced review of Shirky's optimistic assessment in contrast to Nicholas Carr's more pessimistic view (Williams 2010b).

6 According to Wikipedia, Gibson made this observation during a talk about "The Science in Science Fiction" on the NPR show *Talk of the Nation* on November 30, 1999.

7 The meaning of "leadership" has evolved in response to these trends. However, one thing has remained constant over the years: The term "leadership" means many different things to different people. Bennis and Nanus (1985) drive this point home in claiming to have found over 350 definitions of leadership. Similarly, the scholar Peter Northouse (2006) reports that: "In the past 50 years, there have been as many as 65 different classification systems developed to define the dimensions of leadership."

8 Weinberger goes on to note the many risks and costs of clamping down: It introduces inefficiencies, can be demoralizing, can squeeze out innovation, can mask the natural expertise of workers—and is just plain unrealistic.

9 Interview by Deborah Meehan of James McGregor Burns in 2003. Peter Russell quotes the Vietnamese monk Thich Nhat Hanh making a similar point: "The next Buddha will be a sangha [community]." Russell goes on to explain: "The next awakening will come through communal breakthrough rather than the insight of a single being. ... We're going to need that sort of collective thinking to solve some of the problems we're up against."

10 An increasing number of workshops and websites are devoted to this idea. Cf., e.g., The Collective Wisdom Initiative.

11 In a personal communication from April 2009, Kathleen Enright, the president and CEO of Grantmakers for Effective Organizations, describes a trend in her organization's thinking: "We started by looking at how to develop nonprofit leadership as a means of building organizational performance. We weren't focused on individual or collective leadership outside the context of the organization. But understanding that no single organization alone can make significant progress on society's toughest challenges, we are now exploring how to build network and community leadership capacity for social change and problem-solving. That's where our work is moving."

12 Cf. www.inquisitr.com/26835/video-neda-iran-one-life-lost-for-a-greater-cause/.

13 For the past several years, Terri O'Fallon of Pacific Integral has conducted research on the impact of its Generating Transformative Change initiative on participants' level of consciousness. Her as-yet-unpublished

findings suggest that leadership-development programs employing the right combination of challenge and support can have a dramatic impact on participants' level of consciousness (personal communication, Oct. 24, 2010).

14 The "Newtonian paradigm," sometimes known as the mechanistic paradigm, assumed that "things in the environment around humans are more like machines than like life."

15 The term "Millennial" is defined slightly differently by the various authors who use it. For Lancaster and Stillman, it refers to those born between 1982 and 2000; for Meister and Willyerd, to those born between 1977 and 1997 (Meister and Willyerd 2002: 4).

16 Personal communication from Tamara Erickson and Denis Hancock, nGenera, drawing upon U.S. Census Bureau data. Some estimates are more conservative: Meister and Willyerd predict that it will not be until 2014 that Millennials will make up half of the workforce (Meister and Willyerd 2002: 4).

17 Ulrike Reinhard has been particularly helpful on this point. She also alerted us to the important work on networks by Peter Kruse (cf. Kruse n.d., 2010).

18 As mentioned in our second footnote, the best source on this topic is the online journal *Integral Leadership Review*, edited by Russ Volckman.

19 A proven method, based on the work of Doyle and Strauss (1976), is offered by Interaction Associates. Schwartz Associates, another well-regarded source, describes an approach heavily influenced by the Action Science of Chris Argyris in the book *The Skilled Facilitator* (Schwarz 2002). The Institute of Cultural Affairs' Technology of Participation is another well-established method.

20 An organization with a distinctive and well-developed tool kit in this area is The Art of Hosting.

21 After reviewing several successful examples of imaginative solutions to deeply divisive problems that resulted from collaborative networks, Senge et al. observe (Senge et al. 2010: 235): "(N)etwork leaders ... are always asking, 'Who else should we be talking with about this?' In this simple way, existing networks of common interest and concern start to identify themselves. The emerging leaders ... just presented themselves" (Senge et al. 2010: 235).

22 A personal communication from Karen Oshry indicates that such evidence is available and being compiled.

23 Cf. www.powerandsystems.com/ for further information.

24 Andrew Mcfee offers a commentary on each of these 12 in his blog of March 31, 2009.

25 We have seen this work in professional service firms in which the employees have a high degree of autonomy yet respond to economic incentives.

26 A famous example is Lew Gerstner's transformation of IBM (Gerstner 2004).

27 Cf. The World's Most Valuable Brands: Who's Most Engaged? Ranking the Top 100 Global Brands. Thanks to Charlene Li for pointing us to this report and to her own research in a Sept. 30, 2010 webinar on her book *Open Leadership* (2010).

28 Cf. The Global Social Media Check-Up: Insights from the Burson-Marsteller Evidence-Based Communications Group (2009), by Charlene Li.

29 Tapscott is coauthor with Anthony Williams of one of the best books on social media, *Wikinomics: How Mass Collaboration Changes Everything* (2006). He is also currently leading a research initiative with 22 participating companies on the topic of "information technology and competitive advantage" (IT&CA) to assess the potential of evolving business strategies and designs (Buhse and Stamer 2008: 98–120; Saint-Onge 2005).

30 The competitive power of businesses that also have a social bottom line has been dubbed "lethal generosity" (Kanter 2010e).

31 Anklam 2007.

32 Cf. Morgan, Does Collaboration Impact Business Performance?.

33 Levine et al. 2009.

34 Haeckel 1999.

35 Haeckel 1999; Tabscott 2008b: 108.

36 Carr 2008.

37 Annual global mobile data traffic will reach 3.6 exabytes per month by 2014, or an annual run rate of 40 exabytes by 2014, equating to a 39-fold increase between 2009 and 2013. This represents a compound annual growth rate of 108 % (personal communication from Haydn Shaughnessy).

38 Weinberger, in Buhse and Stamer 2008.

39 Personal communication from Tony Adams at the June 2009 nGenera meeting held in San Diego, California.

40 Personal communication from Elmar Husmann on July 7, 2010.

41 Ibid.

42 Wikipedia provides the following history of the phrase: The expression is first recorded as being pronounced by Stewart Brand at the first Hackers' Conference in 1984, in the following context:
   – On the one hand, information wants to be expensive because it's so valuable. The right information in the right place just changes your life. On the other hand, information wants to be free because the cost of getting it out is getting lower and lower all the time. So you have these two fighting against each other.

- Brand's conference remarks are transcribed in the *Whole Earth Review* (May 1985: 49) and a subsequent form of them appears in his *The Media Lab: Inventing the Future at MIT* (1987: 202): "Information wants to be free. Information also wants to be expensive. . . . That tension will not go away."

43 Lawrence Lessig realized the importance of IP many years ago when he founded the "Creative Commons" movement to explicitly tag knowledge on the Web so as to indicate how it can be shared legally (for both free and commercial uses). He realized that Internet technology and social networks would be hampered unless a legal framework was created that could remove some of the barriers caused by modern copyright laws (Lessig 2001).

44 Communication from Peter Holland, a former Cisco employee, now an independent consultant. He went on to add: "Google is a master at both owning intellectual property and in collaborating with people across the globe to their advantage. They use the old model of 'Core Versus Context.' Google openly collaborates for Context (using 'me too' technologies, mostly around open source), which frees up resources and people to focus on Core (new algorithms for searching, new apps so that they can know more about individuals). Other Web-based services—like Facebook, eBay and Amazon—all have much the same strategy."

45 Thomas Gegenhuber, personal communication, Sept. 16, 2010.

46 Quote from Alf Henryk Wulf in a personal communication from Willms Buhse.

47 Willms Buhse, personal communication.

48 April 22, 2009 posting on the Institute of Leadership and Management blog.

49 Company intranet.

50 Information is also available on the BTpedia website.

51 The Technology Adoption Program is a Cisco intranet that is not externally accessible.

52 Snowden (2008) writes of the importance of shifting from "fail-safe design to safe-fail experiments." Shirky argues that "open source" projects—of which the computer software system Linux is a famous example—lower the cost of failure. By relying on peer production, work on projects made accessible through open source code can be highly experimental but at considerably less cost, making it affordable to firms that would otherwise not be able to take such a risk. Failure becomes cheaper than the cost of deciding whether to try something. Shirky concludes from this and similar examples that "services that tolerate failure as a normal case create a kind of value that is simply unreachable by institutions that try to ensure the success of most of their efforts." An example is Meetup, an online service for offering meetings, which "has been consistently able to find (new offerings) without needing to predict their existence in advance and without having to bear the cost of experimentation" (Shirky 2008: 248).

53 "It has always been true that there are a lot more smart people outside any particular company than within it. But there was little to be done about this reality until recently. But that is changing with the help of the Web" (Hagel III, Brown and Davison 2010: 75).

54 One observer notes that Goldcorp had nothing to lose because it owned the site. In addition, he notes, the recognition of the value Goldcorp got has made people warier of doing this for free (personal communication from Michael Chender, founding chair of the Shambhala Institute for Authentic Leadership).

55 Personal communication from Jonathan Imme, July 2010.

56 Tapscott and Williams (2006) provide rich contextual detail about the roots of this example. Cf., as well, Li 2010: 256–257.

57 Cf. Tapscott and Williams 2006: 124–150.

58 "Transparency and authenticity become more than buzzwords because in order for the customer to make intelligent decisions on how they are going to interact with the company and the level of that interaction, they need that visibility and honesty from the company" (Greenberg 2009). Social media are being used to go beyond simply managing customer relations (CRM) to "being able to change how your company does business and improving the user experience." It is not a matter of responding to e-mails or tweets. Rather, Social CRM aims to "actually fix the problems the customers are identifying and collaborating with your customers to help give them what they want. . . . (This is) part of what being a social business is all about" (Morgan 2010).

59 According to Li (2010: 85): "On the eve of the car's launch in December 2009, there had been over six million views of YouTube videos, 740,000 views of Flickr photos and 3.7 million Twitter impressions."

60 Personal communication from Dennis Hancock, researcher for nGenera (now Moxie Software) at the July 27–28, 2010 nGenera meeting held in Carlsbad, California.

61 Relationships in which customers are both producers and consumers (cf. Tapscott and Williams 2006).

62 Spy is an intranet that is not accessible outside Best Buy.

63 This portrait draws on multiple sources. Former Cisco employee Peter Holland, now an independent consultant, whom we tapped as part of our pool of experts, first called our attention to the company as a noteworthy example. We also drew upon the book that Cisco has published to document its experiment and offer lessons to others (Sankar and Bouchard 2009). A consultant/executive team provided an inside look that was originally proprietary but has since been made public (Tapscott 2008b). Finally, we had the benefit of Charlene Li's case study in her recent book (2010).

64 Personal communication from nGenera's Tamara Erickson, June 23, 2010.

65 Personal communication from Ulrike Reinhard, July 2010.

66 CoreMedia was the winner in the major enterprise category, chosen jointly by AT Kearney and the German business magazine *Wirtschaftswoche*, in recognition of its achievements in innovation management (cf. Buhse and Stamer 2008).

67 Willms Buhse oversaw the creation of a case study based on CoreMedia that was used at Harvard University.

68 This case profile draws heavily from—and retains much of the original language of—a detailed and excellent description in Hagel III, Brown and Davison 2010.

69 Beth Kanter offered a comment on this section (personal communication, Oct. 25, 2010): "We've gotten past early adopters, and more nonprofits are exploring. I no longer talk to groups of nonprofits and get the same level of skepticism as I did in the past. They've gone from 'Do we have to?' to 'How do we get started in a strategic way?'"

70 Levy's first blog post, quoted in Li 2010: 27.

71 Personal communication from Thomas Moroz, codirector of KARL, August 2009.

72 Cf. https://nitrogen.packard.org.

73 Just as we were finalizing this report, we received suggestions from Beth Kanter based on her judgment of what the best examples were. She cited Case Foundation's America's Giving Challenge and referred to a blog she has written on a 2010 SXSW conference panel (Kanter 2010a).

74 Cf. www.tcfn-cfc.ca/wp-content/uploads/2009/07/susan-mernit-the-social-media-toolbox1.pdf.

75 Some readers may find it useful to know that "as Acumen progressed in this work, major partners such as Google and Salesforce.com joined in and began the push to create measures and tracking systems that could be used by other organizations, as well as to enable it to raise more investment dollars. Doing so required the development of a shared taxonomy of outcomes and of systems that could track information within a single organization as well as feed (it) into a common database. Thus was born the Pulse platform—a software system for tracking outcome measures. The Impact Reporting and Investing Standards (IRIS), a shared taxonomy of outcome definitions, is currently being launched alongside the Pulse platform" (Bernholz, Skloot and Varela 2010: 27).

76 Bernholz's elaboration is worth noting (ibid.: 28): "Recognition of the value of the end-user experience is inherent in the process. Students have access to the information as well as other resources that might help them improve their schools. This type of evaluation turns subjects into actors. It changes dynamics at every level—when information is collected, from whom, how it's used and who can analyze it—at a cost that is negligible when compared to traditional approaches."

77 "FutureChallenges—Changing Leadership Requirements in entering the World of Web 2.0," with Andreas Esche, Henrik Scheller and Ole Wintermann.

78 Personal communication from Bertelsmann Stiftung Senior Project Manager Ole Wintermann, August 2010.

79 Cf. http://en.wikipedia.org/wiki/Bureaucracy.

80 Not only does this domain lack the tangible performance measures available in the business sector (e.g., profit margins, stock prices, market share), it also lacks the more specific performance measures of social-sector organizations as well (e.g., attainment of program goals, fundraising benchmarks). Government agencies also have inherently more ambiguous measures of performance because they simultaneously pursue multiple, noneconomic goals (van Slyke and Alexander 2006: 364).

81 Beth Noveck is professor of law and director of the Institute for Information Law and Policy at New York Law School and McClatchy Visiting Professor of Communication at Stanford University.

82 Cf. http://gov20australia.ning.com/profiles/blogs/australian-government-responds.

83 Cf., e.g., "Feds Join Twitter Revolution," a March 9, 2009 posting on *NetworkWorld* (www.networkworld.com/news/2009/030909-feds-twitter.html).

84 Cf. this blog post on government: www.theregister.co.uk/2009/04/03/google°n_washington/

85 Cf., e.g., "White House to Host Dialogue Solutions for Recovery.gov," an April 24, 2009, posting on the website www.goloop.com.

86 Intellipedia is an intranet to which—for security reasons—there is no link.

87 In a memo to six federal agencies, Kathleen Sebelius and HHS director Peter Orszag wrote that this new task force would replace the existing health IT interagency group: "This legacy structure is not a good fit for the new environment that includes a statutory Office of the National Coordinator with greatly enhanced policy-making responsibilities, two new Federal Advisory Committees (FACAs), increased congressional engagement and attention from a diverse body of interests in the private and public sectors." The task force will include the departments of Agriculture, Commerce, Defense and Veterans Affairs, the Social Security Administration, the Office of Personnel Management, the federal chief information officer and the federal chief technology officer. The memo continues: "The purpose of the HIT Task Force will be to assist with policy development, coordination and implementation of Federal HIT activities, as well as to improve trans-

parency of Federal government activities related to HIT and communication among Federal agencies as they execute Federal HIT policy." Cf. also: http://ahier.blogspot.com/2010/02/federal-health-it-task-force.html.

88 There is no single URL for this project.

89 Cf. Wikipedia under "Open Source Governance."

90 Ibid.

91 All of these profiles are taken from a personal communication from Thomas Gegenhuber, an advocate/researcher/writer on Web 2.0 living in Linz, Austria.

92 Freie Netze. Freies Wissen.

93 Thanks to nGenera's Nauman Haque for alerting me to the case.

94 Personal communication from Thomas Gegenhuber, September 16, 2010.

95 Personal communication from Ole Wintermann, August 2010.

96 Personal communication from Thomas Gegenhuber, September 16, 2010.

97 There seems to be no single URL for this program.

98 Kelty provides a rich history of the interwoven movements for free software, software and open source code in *Two Bits* (Kelty 2008).

99 A new form of activism—"micro-movements"—now enables individuals or small groups to create a "movement in a silo" by connecting with people through the Web and enabling followers to easily connect with one another. Many of the initiatives that enable individuals (as free agents) to act as micro-volunteers are the product of individuals acting as free agents in a more entrepreneurial sense. Chapter 3 has already cited a number of such examples.

100 This "nudge" approach is described at length in Thaler and Sunstein 2009.

101 Learning histories and reports from the Sustainable Food Lab can be found at www.sustainablefoodlab.org. A description of some of its activities, including learning journeys, can be found in Senge et al. 2010.

102 For reflections on the Bhavishya Alliance, cf. Hassan and Bojer 2005.

103 This term is being used by multiple, apparently independent actors, which attests to the movement's growing momentum (e.g., the Conscious Capitalism Alliance, sponsor of the C3 conferences, and Bentley University, sponsor of a different set of "C3" conferences on Conceptualizing Conscious Capitalism).

104 There is, of course, a chicken-and-egg problem here. One's strategy should reflect fundamental goals. But those goals should reflect knowledge of the environment. Learning about the environment could suggest the need to adjust basic goals. But the best way to start is to simply acknowledge where you are and move on from there.

105 Marilyn Darling, expert on after-action reviews, tells us that (personal communication, June 10, 2010): "The gold standard of a learning organization is OPFOR ('Opposing Force'), whose job it is to represent the enemy in simulated war games … (and) to be the 'thinking and uncooperative enemy' to brigades of US soldiers that are preparing to deploy to Iraq or Afghanistan. What they have learned to do through sheer discipline is to get good at predicting what the enemy will do and planning how to respond."

106 *McKinsey Global Survey 2009: How Companies Are Benefiting from Web 2.0*, cited in Stamer 2009.

107 CoreMedia commissioned Berlecon Research to conduct a representative survey in the summer of 2007 that would cover the usage and appraisal of Web 2.0 within German corporations. The survey questioned 156 senior management personnel employed at companies in "knowledge-intensive" industries with more than 100 employees. These industries were chosen because of their pressing need for efficient knowledge and information management, which also means that the potential for introducing Enterprise 2.0 in these companies is especially high. Interviews were computer-supported, over-the-phone dialogues in which management personnel from R&D, marketing, public relations and human resources were asked their opinions on the following topics: What are the current key challenges within the company regarding teamwork and the exchange of knowledge and information? How have these challenges evolved over the last few years? What is the company position with regard to the relevance and benefit of Web 2.0 applications? (Buhse, in Buhse and Stamer 2008:141).

108 Li also points out that Apple is open in many ways, for example, in its popular user forums.

109 There are of course many ways of framing the stages of the adoption of Web 2.0. Tapscott and Williams (2006) write of a "staged plateau," which assumes that the organization has made a decision to adopt Web 2.0 practices. In their words, this staged plateau has the following levels:

1. Beachhead: Launch strategic pilots. Employ disposable, easy to use, inexpensive technologies. This has no significant risk. It requires the organization to first define the need and garner support for the next phase.

2. Basecamp: Prove the concept is scalable. Move from "push" to "pull," where IT no longer has to push collaborative efforts. Business groups see value of the wiki workplace and allocate budgets.

3. Focus on continuous improvements, enhancing, extending. A broad governance structure and architectural strategy are required.

110 The author went from theoretical to hands-on knowledge of Web 2.0 by enrolling in a course, "Social Media Jedi" (offered by Michael Margolis), which turned out to be highly useful.

111 Personal communication, Oct. 25, 2010.

112 The courses were offered by Blue Oxen Associates and Get Storied.

113 Personal communication from Ole Wintermann, Senior Project Manager, Bertelsmann Stiftung.

114 E.g., JetBlue, with its JetBlue University Blog (cf. Meister and Willyerd 2010: 220).

115 Bell Canada did this to increase customer satisfaction and create an Innovation Jam (ibid.: 144).

116 Such as: "It's only the Millennials that will use this at work," "If we build it employees will come," "Employees will inevitably share company secrets," "It's a security risk," "It's just a fad," or "It will mean decreased productivity due to online networking" (ibid.: 145).

117 The survey was conducted by Robert Half Technology. It was cited in Li 2010: 277 (footnote 1).

118 Personal communication, Oct. 25, 2010.

119 Beth Kanter gives six dozen examples of CEOs (mostly of nonprofits) who have a blog or use Twitter. She also offers seven tips (Kanter 2010c).

120 Personal communication, May 19, 2010.

121 "The typical problem here is lack of a persistent internal communications/marketing campaign to carry usage to a critical mass after the initial enthusiasm." Personal communication from Michael Chender, founding chair of the Shambhala Institute Authentic Leadership program.

122 Personal communications from KARL codirectors Thomas Moroz and Yalan Teng.

123 Personal communication from a source who prefers to remain anonymous.

124 Personal communication, July 2010.

125 Cf. e.g., Kanter 2010f.

126 William Powers puts the interpenetration of work time and personal time in historical context, going back to the Roman philosopher Seneca. Drawing on the experience of his family, he makes the case for regular "Internet sabbaticals" (Winer 2010).

127 Recent hoopla surrounding Google Buzz and the management of Facebook privacy settings may be inconsequential compared to the capacity of HTML 5 to quietly gather data on the preferences of Internet users (Vega 2010).

128 This is especially true in universities, particularly in the humanities.

129 Cf. Kanter 2009b.

130 Counterbalancing all of the positive benefits of the Web are the many ways in which it can be used for illegal, immoral or socially destructive purposes. To take an extreme example from a value-neutral perspective, al-Qaeda represents a best-practice use of Web-enabled networking (Thompson 2006). The Web has also been the most effective vehicle for recruiting terrorists.

131 The terms "Web" and "Internet" are often used interchangeably. We distinguish between them as follows: The "Internet" is the system of interlinked computers that provides a technological foundation of hardware and software on which the pages of the "Web" appear. Our focus in this study is on the Web.

132 According to Wikipedia: "The World Wide Web (commonly abbreviated as 'the Web') is a system of interlinked hypertext documents accessed via the Internet. With a Web browser, one can view Web pages that may contain text, images, videos and other multimedia and navigate between them using hyperlinks. Using concepts from earlier hypertext systems, the World Wide Web was started in 1989 by the English physicist Sir Tim Berners-Lee, now the Director of the World Wide Web Consortium, and later by Robert Cailliau, a Belgian computer scientist, while both were working at CERN, in Geneva, Switzerland. In 1990, they proposed building a 'web of nodes' storing 'hypertext pages' viewed by 'browsers' on a network, and (they) released that web in December. Connected by the existing Internet, other websites were created around the world, adding international standards for domain names and the HTML language."

133 Practices diverge regarding Internet capitalization conventions, with good arguments on both sides. We have decided to follow the convention of capitalizing both "Web" and "Internet."

133 These definitions draw heavily on Brotherton et al. 2008 (esp. pp. 41–42) and McKinsey & Co. 2007 (p. 6).

134 "Crowdsourcing" is a term coined by the writer Jeff Howe in a 2006 article in *Wired* magazine to describe the phenomenon of using large, dispersed groups of amateurs networked through the Web to do work that was previously performed by solitary experts or units within larger institutions.

135 McAfee has acknowledged that, although he thought he had coined the term, it was actually used by the British Internet consultant Stuart Eccles a month before he used it in March 2006 (McAfee 2006b).

136 Personal communication from Jonathan Hooper of the Open Society Institute, April 24, 2009.

The Authors

Grady McGonagill, Ed.D, is principal of McGonagill Associates, an organizational consulting and management development firm which since 1983 has specialized in building capacity for learning and change. He has distinctive expertise in leadership development, developing a culture supportive of leadership and learning, and executive coaching. He has a master's degree from Stanford University and a doctorate from Harvard University. Grady is a contributor to the *Fifth Discipline Fieldbook*, edited by Peter Senge et al. (New York: Doubleday, 1994) and the author of a chapter in *Executive Coaching*, edited by C. Fitzgerald and J. Berger (San Francisco: Davies Black Publishing, 2002). www.mcgonagill-consulting.com

Tina Doerffer, MPA, holds a law degree from the Humboldt University of Berlin and a Master's of Public Administration from the Harvard Kennedy School of Government. She is a member of the German Bar Association and lectures at the Center for Public Leadership, Harvard University, among other venues. As a project manager at the Bertelsmann Stiftung she currently heads the Bertelsmann Stiftung Leadership Series in the Corporate Cultures in a Globalized World program. www.bertelsmann-stiftung.de/leadership

Maria Stippler, Sadie Moore,
Seth Rosenthal, Tina Doerffer

Leadership
Approaches—Developments—Trends
Bertelsmann Stiftung Leadership Series

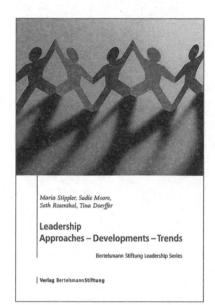

2011
96 pages, softbound
€ 16,00
ISBN 978-3-86793-323-0
Available as an ebook

Everyone is talking about leadership but what are the common approaches, camps, and theories? What is current, what are the new classics, and what is obsolete?

The crisis and the latest Web 2.0 developments have not rendered the topic any less relevant. Which school of thought is closest to yours? Which approach informs your actions as a manager? The five-part "Leadership" reader, with its overview of approaches, developments and trends, provides references and guidance to help you anchor your own point of view. Our aim is to provide support to you in your daily, practical work with your executive board, colleagues and employees, and to contribute to the discussion of leadership in Germany.

Read Part 1: Earliest Theories, Part 2: Systemic Leadership, Part 3: Leadership as a Relational Phenomenon, Transformational Leadership, Values and Ethics, Part 4: Motivation, Power and Psyche and Part 5: Leadership Today. The publication is available as an ebook.